Physiological Psychology

The INSTANT NOTES series

Series Editor: B.D. Hames, School of Biochemistry and Molecular Biology, University of Leeds, Leeds, UK

Animal Biology 2nd edition
Ecology 2nd edition
Genetics 2nd edition
Microbiology 2nd edition
Chemistry for Biologists 2nd edition
Immunology 2nd edition
Biochemistry 2nd edition
Molecular Biology 2nd edition
Neuroscience
Developmental Biology
Plant Biology
Bioinformatics

Chemistry series
Consulting Editor: Howard Stanbury

Organic Chemistry 2nd edition
Inorganic Chemistry 2nd edition
Physical Chemistry
Medicinal Chemistry
Analytical Chemistry

Psychology series
Sub-series Editor: Hugh Wagner, Dept of Psychology, University of Central Lancashire, Preston, UK

Psychology
Cognitive Psychology
Physiological Psychology

Forthcoming title
Sports and Exercise Psychology

Physiological Psychology

H. Wagner

Department of Psychology, University of Central Lancashire,
Harrington Building, Preston, UK

and

K. Silber

Psychology Department, Staffordshire University,
College Road, Stoke-on-Trent, UK

BIOS Scientific Publishers
Taylor & Francis Group

LONDON AND NEW YORK

© Garland Science/BIOS Scientific Publishers, 2004

First published 2004

A CIP catalogue record for this book is available from the British Library.

ISBN 1 85996 2033

Garland Science/BIOS Scientific Publishers
4 Park Square, Milton Park,
Abingdon, Oxon OX14 4RN, UK and

29 West 35th Street, New York,
NY 10001–2299, USA
World Wide Web home page: www.bios.co.uk

Garland Science/BIOS Scientific Publishers is a member of the Taylor & Francis Group

Distributed in the USA by
Fulfilment Center
Taylor & Francis
10650 Toebben Drive
Independence, KY 41051, USA
Toll Free Tel.: +1 800 634 7064; E-mail: taylorandfrancis@thomsonlearning.com

Distributed in Canada by
Taylor & Francis
74 Rolark Drive
Scarborough, Ontario M1R 4G2, Canada
Toll Free Tel.: +1 877 226 2237; E-mail: tal_fran@istar.ca

Distributed in the rest of the world by
Thomson Publishing Services
Cheriton House
North Way
Andover, Hampshire SP10 5BE, UK
Tel.: +44 (0)1264 332424; E-mail: salesorder.tandf@thomsonpublishingservices.co.uk

Library of Congress Cataloging-in-Publication Data

Wagner, Hugh L.
  Physiological psychology / H. Wagner, K. Silber.
      p. cm. — (Instant notes)
  Includes bibliographical references.
  ISBN 1-85996-203-3
    1. Psychophysiology—Outlines, syllabi, etc. I. Silber, Kevin, 1959– II. Title. III. Series: Instant notes series.

QP360.W335 2004
612.8—dc22   2004003212

Production Editors: Harriet Milles and Catherine Jones
Typeset by Phoenix Photosetting, Chatham, Kent, UK
Printed by Biddles Ltd, Kings Lynn, UK, www.biddles.co.uk

CONTENTS

ABBREVIATIONS

5HT	5-hydroxy-tryptamine (serotonin)	LH	luteinizing hormone
ACh	acetylcholine	LPN	lateral parabrachial nucleus
AChE	acetylcholinesterase	LSD	lysergic acid diethylamide
ACTH	adrenocorticotropic hormone	LTP	long-term potentiation
ADH	antidiuretic hormone (vasopressin)	MAOI	monoamine oxidase inhibitor
AMH	anti-Müllerian hormone	MDMA	methylenedioxymethamphetamine (ecstasy)
AN	anorexia nervosa		
ANS	autonomic nervous system	MP	muramyl peptide
BN	bulimia nervosa	MPA	medial preoptic area
BNST	bed nucleus of the stria terminalis	MPN	median preoptic nucleus
BRAC	basic rest–activity cycle	MPRF	medial pontine reticular formation
Ca^{2+}	calcium ions	MSG	monosodium glutamate
CAT scan	computerized axial tomography	Na^+	sodium ions
CCK	cholecystokinin	NMDA	N-methyl-D-aspartate (receptors)
CNS	central nervous system	NPY	neuropeptide Y
CR	conditioned response	NST	nucleus of the solitary tract
CRF	corticotropin releasing factor	OVLT	organum vasculosum of the lamina terminalis
CRH	corticotropin-releasing hormone		
CS	conditioned stimulus	PAG	periaqueductal gray
DMTN	dorso-medial nucleus of the thalamus	PET scan	positron emission tomography
		PIH	prolactin-release inhibiting hormone
DNA	deoxyribonucleic acid		
ED50	median effective dose	POAH	preoptic area and anterior hypothalamus
EE	expressed emotion		
EEG	electroencephalogram/ electroencephalography	PRH	prolactin-releasing hormone
		PTSD	posttraumatic stress disorder
EPSP	excitatory postsynaptic potential	PVN	paraventricular nucleus
fMRI	functional magnetic resonance imaging	RAS	reticular activating system
		REM sleep	rapid eye movement sleep
FSH	follicle-stimulating hormone	RSBD	REM sleep behavior disorder
GABA	gamma-aminobutyric acid	SAD	seasonal affective disorder
GAS	general adaptation syndrome	SCN	suprachiasmatic nucleus
GH	growth hormone (somatotropin)	SDN	sexually dimorphic nucleus
GHB	gamma-hydroxybutyrate	SFO	subfornical organ
GI	gastrointestinal	SNRI	selective norepinephrine reuptake inhibitor
GnRH	gonadotropin-releasing hormone		
GRH	growth-hormone-releasing hormone (somatocrinin)	SSRI	selective serotonin reuptake inhibitor
		SW sleep	slow-wave sleep
HSE	Herpes simplex encephalitis	TABP	Type A behavior pattern
INAH-3	third interstitial nucleus of the anterior hypothalamus	THC	tetrahydrocannabinol
		TRH	thyrotropin-releasing hormone
IPSP	inhibitory postsynaptic potential	UCR	unconditioned response
K^+	potassium ions	UCS	unconditioned stimulus
LD50	median lethal dose	VMH	ventromedial hypothalamus
LGN	lateral geniculate nucleus	WKS	Wernicke–Korsakoff syndrome
LH	lateral hypothalamus		

PREFACE

evol -
continuity

1. Biology
3. Nervous System
- anatomy &
- physiol
) peripherl
Nersystm
F. Endocrine
System
- Drugs
- Vision
+ Hearing

Psychology is the study of behavior and mental processes. This study cannot be complete without considering the biological context in which behavior and mental processes occur. In the broadest sense, this means that we consider human beings to be the result of evolution. This book is particularly concerned with the physiological processes upon which all behavior and mental processes depend. This subject is known by various names, including *physiological psychology*, *psychobiology*, *biological psychology*, *biological bases of behavior*, *physiology of behavior*, and *biopsychology*. Our starting point is that we cannot hope to understand psychological phenomena fully without examining their physiological basis. Evolution implies continuity of anatomical structures and physiological processes with other species. This means that we can learn much by studying physiological processes in other species.

This book is intended as a study guide or revision aid in physiological psychology. We present the key ideas and facts in this field of study in 18 sections, divided into a total of 63 topics. Each topic has a Key Notes panel which summarizes concisely the main points covered in the main text of the topic. The main text includes illustrations that are mostly simple line drawings. These are mainly limited to figures which will assist the student in understanding the material, and which should be easy for the student to reproduce. Each topic stands alone, but, especially for later sections, understanding will be helped by referring back to earlier topics. For this reason we provide a list of related topics for each topic.

The contents of the book have been selected on the basis of our experience of many years of teaching physiological psychology in universities in the UK. We have also considered the syllabus of courses taught both here and abroad. In the first part of the volume we concentrate on the underlying biological, anatomical and physiological structures and principles. Section A examines the biological background and methodology of physiological psychology. This includes consideration of genetics and evolution. It is worth emphasizing here an axiom that guides us and should guide the student. Anatomical structures and physiological processes are largely determined by genetics. This implies that, in turn, psychological processes are influenced by genes. However, it is impossible to say that behavior is *caused* by genes. Equally, we cannot say that behavior is *caused* by environmental factors. We reject both of these extremes of the *nature–nurture* debate in favor of a view that holds that behavior and mental processes result from an interaction between genetic and environmental factors.

The next four sections cover the 'hardware' on which behavior depends. Section B describes the components of the nervous system, and how information is transmitted along and between neurons. Section C provides an account of the basic anatomy and physiology of the body's central 'processor': the central nervous system. This includes an overview of the effects of brain damage. This is followed in Section D by a description of one of the parallel effector systems: the peripheral nervous system. This has two main components: the somatic and the autonomic nervous systems. The other effector system, the endocrine system, is covered in Section E. Here we look at the nature of hormones and how they act on different tissues, and look at the specific endocrine glands that are of importance in psychological phenomena. The central and peripheral nervous systems, and the endocrine system, are all affected by various drugs. The ways in which drugs act on the body, and the effects they have, are covered in Section F. We examine drugs that have direct effects on mental states (psychoactive drugs), and the issues of drug tolerance and dependence.

The next three sections look at how the body obtains information about the environment. Probably our most important sense is vision, which is the subject of Section G. This follows visual information from its reception by the eye to its processing in the brain. Section H is devoted to the next major sensory system: audition. This follows auditory information from its reception in the ear to its representation in the central nervous system, and describes how the important types of information contained in sound are extracted from the auditory signal. The other senses, the somatosensory senses of movement, balance, touch, and pain, and the chemical senses of gustation (taste) and olfaction (smell) are the subject of Section I.

Having obtained information from the environment, and in order to seek it, animals behave by moving around in the environment. Section J examines the components of the motor system: muscles, reflexes, motor pathways from the central nervous system, and motor areas of the central nervous system. One characteristic of behavior is that it is not constant throughout the day and night. Section K concerns rhythmic variations in activity, and how they are controlled. This includes the most obvious day–night variation: sleeping and waking. Topics in this section consider the nature, control and possible functions of sleep and rhythmic activity, and the effects of their disruption.

Another major characteristic of behavior is that it is motivated. Some of the ways in which it is motivated are in the pursuit of biological needs. In Section L we introduce the idea that much behavior serves to maintain the internal environment (homeostasis). We consider the maintenance of body temperature and fluid regulation. Section M examines how far such homeostatic processes can explain eating and body weight control. In doing this, we consider the basic facts about digestion and energy storage and mobilization, physiological control mechanisms, and disorders of weight control. Section N looks at the non-homeostatic biological motivations of sex and reproduction. We start by reviewing the nature and origins of the differentiation of males and females, and consider in turn hormonal and neural mechanisms controlling or influencing sexual and reproductive behavior, including parental behavior and sexual orientation. In Section O we look at the nature, origins, and communication of the emotions that accompany motivated behavior. We consider also stress, psychosomatic illness, and aggression.

Section P looks at another way in which animals do not simply respond to environmental stimuli: they use learning and memory. We look at types of learning and memory, and their underlying neural mechanisms. Communication is important in all species. In Section Q we examine language and its disorders, and their neural basis. Finally, Section R is concerned with three major classes of psychological abnormality: schizophrenia, mood (or affective) disorders and anxiety disorders.

The aim of this book, as with others in the *Instant Notes* series, is to present the core material of its subject area. It should not be expected to replace the expanded coverage of conventional textbooks, and we refer to the best of these in the Further Reading section that follows Section R. We also direct you to more specialized sources relevant to each section so that you can pursue these in more detail if you so wish.

A1 PSYCHOLOGY AND BIOLOGY

Key Notes

The biological context of psychology	Psychology is a biological science: humans are products of evolution, and mental processes and behavior are dependent on the body. Evolutionary psychology examines the evolution of human behavior. Applied to social behavior this is sociobiology. Physiological psychology studies the physiological mechanisms underlying psychological processes.
Reductionism	Explaining mental processes and behavior in physiological terms is reductionism. While some argue that reductionism is flawed, few physiological psychologists would argue that theirs is the only useful approach. Reductionism does not imply that one particular level of analysis is 'correct'. We can choose levels of analysis that are appropriate for our purposes.
Human evolution	Darwin proposed that evolution proceeds by a process of natural selection. Natural selection applies to individuals, not groups. It is the fittest individual who survives, not the fittest species. Changes in fitness come from mutations in genes. Most mutations are harmful, and the individuals carrying them are less likely to reproduce. Advantageous mutations can provide a selective advantage. Eventually, accumulated advantageous mutations make the individual so different from its ancestors that a new species emerges.
Genetics and the 'nature–nurture' issue	Each gene carries the code for the production of a particular molecule. These molecules are the basis of cell and tissue development. We have 23 pairs of chromosomes in the nucleus of most of our cells, and inherit one of each pair from each parent. This produces genetic diversity in the offspring. Genes on the same chromosome are not always inherited together. The genetic makeup of each individual organism is its genotype. The extent to which a gene exerts its effect depends on environmental influences, resulting in the phenotype. In discussing the old 'nature–nurture problem' people often took either/or positions on the roles of genetic and environmental factors. Genes do not directly determine behaviors, but affect biochemical processes involved in behavior.
Related topic	Methods of research (A2)

The biological context of psychology	Psychology is the study of behavior and mental processes. Our chief concern is with human psychology. In at least two ways, psychology is a biological science:

- humans are products of **evolution**. This continuity with other species means that we can make meaningful comparisons with the behavior of other species. For example, principles of learning have been studied in experiments on rats. It also means that psychology has to consider how evolution has shaped human behavior;

● mental processes and behavior are dependent on the **nervous system** (see Sections C and D) and the **endocrine system** (see Section E).

Biological psychology (or **psychobiology**) includes all approaches to psychology that place it in its biological context, examining the biological bases of behavior. One such approach is the examination of the evolution of human behavior, in a field known as **evolutionary psychology**. This discipline tries to extend to human behavior principles from studies of other species. When this approach is applied to social behavior the field is called **sociobiology**. Sociobiology applies principles of evolution, predominately Darwinian, to the study of social behavior. A key concept is the **selfish gene** (Dawkins, 1989). In this view, animal behavior is interpreted as having the goal of increasing the proportion of the individual's genes in the next generation.

This book is concerned with **physiological psychology**. The biological approaches mentioned above are all involved with studying how human beings fit into the *broader* biological context. Physiological psychology takes a narrower approach. It involves the study of the physiological mechanisms, especially those of the nervous and endocrine systems, underlying psychological processes. Thus, for example, we will in this book examine the neural mechanisms involved in a range of emotional and motivational states. However, it is important to note that *all* the systems of the body (including those concerned with immune function, digestion, reproduction, circulation, respiration, etc.) work as an integrated whole. It is in the study of these mechanisms that psychology relies most heavily on assumptions such as similarity of mechanisms across species, and evolutionary continuity. Within physiological psychology **neuropsychology** is the study of the neural mechanisms, especially those in the cerebral cortex, which underlie psychological, particularly cognitive, processes.

Reductionism

The aim of any science is *explanation*. Physiological psychology has the aim of explaining human mental processes and behavior in physiological terms. The practice of explaining complex events in terms of simpler ones is known as **reductionism**. For example, memorizing a telephone number can be described at several levels. At the psychological level, we could formulate rules describing how such memories are formed and maintained in the memory. A cognitive psychologist might try to explain it in terms of aspects of a modular memory system. A physiological psychologist might examine the neural processes that form and store the memory, perhaps trying to identify the neural basis of the modules described by the cognitive psychologist. A biochemist could try to describe the chemical changes underlying the neural processes. A physicist might explain these chemical changes in terms of the forces interacting at the molecular and submolecular level. Each of these attempts at explanation is an example of reduction: reduction to an underlying, and in some sense simpler, level.

Some psychologists argue that reductionism is fundamentally flawed. For them, psychology is, or should be, capable of both describing and explaining phenomena at their own level. But this is to misunderstand several aspects of reductionism in general, and the nature of physiological psychology in particular:

- few physiological psychologists would argue that theirs is the only approach worth taking. Without the findings of those who study psychology at the behavioral or mental level, the physiological approach would be empty;

- most physiological psychologists use principles from different levels in trying to explain particular behaviors. For example, none would say that a particular behavior or condition is 'caused by' genes. Rather, it is accepted that genes act in interaction with physical and social environmental factors. Similar arguments apply to relationships amongst physiological and psychological levels of description. This approach might be termed **interactionist**;

- the physiological approach often suggests psychological processes. For example, the study of localized brain damage has suggested how language is organized (see Topic Q1);

- reductionism does not imply that one particular level of analysis is 'correct'. We can choose a level of analysis, or combination of levels, that is most appropriate for our purposes. For example, it would be little help to a therapist to try to describe depressive illness at the level of subatomic physics. However, approaches at the social, psychological, physiological and biochemical levels might all be useful, separately or together.

In fact, as suggested in the last point, anyone who might want to limit psychology to the study of the whole individual is missing another 'higher' level of analysis. Everyone is influenced by the family and other social groups to which they belong. Since we live in social groups, we must consider the social context of human mental processes and behavior, and this makes psychology a **social science**, as well as a biological one.

Human evolution
In his book *The Origin of Species*, Charles Darwin (1859) proposed that evolution proceeds mainly by a process of **natural selection**: the 'survival of the fittest'. However, Darwin was not the first to suggest that we have evolved from other creatures. Naturalists had for a long time classified animals and plants into groups of similar species; a process called **taxonomy**. Taxonomy places species into a 'tree' related to one another in different degrees by similarities of their structures. The concept of evolution made sense of these relationships amongst species, now known as the **phylogenetic tree** (see *Fig. 1*). Some animals are more similar because they diverged from a common ancestor more recently in evolutionary history.

Natural selection applies to individuals, not groups. That is, it is the fittest *individual* who survives, not the fittest *species*. Changes in fitness come from **mutations**: changes in the genes resulting, for example, from natural background radiation. Most mutations are harmful, and the individuals carrying them are less likely to reproduce, perhaps dying before birth. Advantageous mutations can increase fitness, or provide a **selective advantage**, in many different ways. They can, for example, provide disease resistance, improve protective coloration, enhance attractiveness to the opposite sex, improve sensory acuity, permit the digestion of a different food, or (by way of neural changes) improve behavioral programs that permit more effective behaviors involved in hunting, mating and the like. Eventually, accumulated advantageous mutations make the individual so different from its ancestors that a new species emerges.

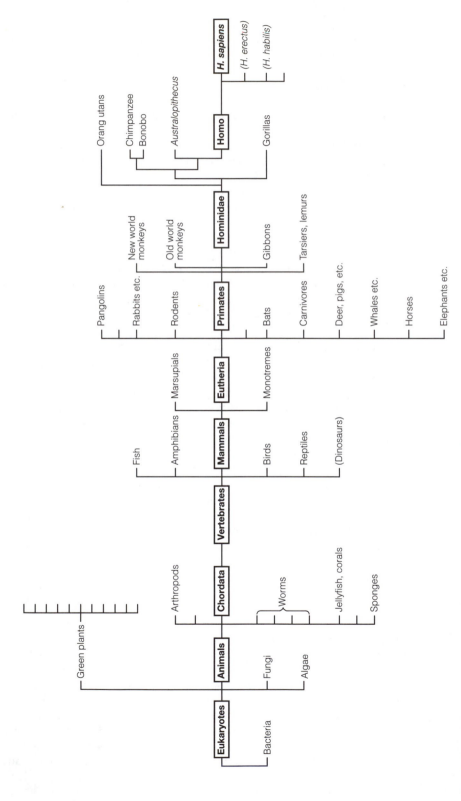

Fig. 1. Simplified diagram of part of the phylogenetic tree of humans. Most branches and some nodes in the tree have been omitted.

All present day species are the latest step in their line of descent. We are *not* descended from any other existing species. Our line of descent diverged from that of the chimpanzees at a common ancestor some 5–7 million years ago. Nevertheless, we share most of our genes, and much of our physiology, with other species. This makes it possible to gain an understanding of human psychology, and in particular physiological psychology, by studying other species.

Genetics and the 'nature–nurture' issue

All of the tissues in our bodies derive from **genes** that we inherit from our parents. Each gene carries the code for the production of a particular protein or other molecule. These molecules are directly or indirectly the basis of cell and tissue development. Each gene is part of a very large molecule called **deoxyribonucleic acid** (**DNA**). Each DNA molecule is looped and bound to a central matrix to form a **chromosome**. We have 23 pairs of chromosomes in the nucleus of most of our cells, and inherit one of each pair from each parent through the sperm and egg cells. This separation of chromosomes so that each egg or sperm carries half of the nuclear DNA of each parent produces genetic diversity in the offspring. This diversity is increased by the process of **recombination**, or **crossing over**. Before the chromosomes in sperm and egg cells split, varying amounts of the DNA from each pair of chromosomes are exchanged with the paired chromosome. The result of crossing over is that genes on the same chromosome are not always inherited together. The closer two genes are to each other physically on a chromosome, the more likely they are to be inherited together. There is also DNA in the **mitochondria** (structures in which energy is produced) of each cell. This mitochondrial DNA is passed intact from the mother to every one of her offspring.

The genetic makeup of each individual organism is called its **genotype**. However, it is important to remember that the extent to which a gene exerts its effect depends on environmental influences. The resulting characteristics of the individual are known as its **phenotype**. For example, genes that specify the physical characteristic of height will only ensure that you are tall if you are adequately nourished during growth. For a large part of the 20th century, psychologists debated the roles of innate and environmental factors on psychological processes: the **nature–nurture problem**. Many took extreme positions, dictated more by ideology than by science. It has become clear to most scientists that for most psychological processes, genetic and environmental factors interact to produce the behavioral phenotype.

Molecular biology can identify the genetic makeup of a particular organism, and can find the sites of the genes which code for particular molecules, and thereby control the development of particular structures. In this way, behavioral genetics should be able to locate genes significant for the production of particular behaviors or psychological traits. We stress that genes do not determine behaviors. Genes produce molecules that influence structural characteristics. These characteristics, such as the development of particular brain structures or receptors on cells, in turn influence behavior.

Biologists can now define phylogenetic relationships in terms of the similarity of the DNA of different species. We share more than 98% of the DNA of chimpanzees, and slightly less than 98% of that of gorillas. While this common inheritance establishes phylogenetic continuity, the 1.6% or so of our DNA which we *do not* share with other species is clearly extremely important, and defines our differences from the other primates. Primates, and in particular human beings, are characterized by the evolution of progressively larger brain size, in relation to body size, especially of the **cerebral hemispheres** (see Topic C1). This process is known as **encephalization**. Encephalization and the accompanying increase in brain complexity provide enhanced individual adaptability, and this increases along the phylogenetic scale to the human brain.

A2 METHODS OF RESEARCH

Key Notes

Lesion studies	Brain lesions may be produced by injury or disease, or produced surgically. Lesions resulting from accident or disease are rarely localized. Experimental lesions permit much more localized lesions to be studied. To assist the production of lesions in precisely defined parts of the brain a stereotaxic apparatus is used. The exact location of lesions can usually only be described *post mortem*. Lesioned areas may have general rather than specific functions. We need to be careful that the area destroyed controls the target behavior, and not some other essential behavior.
Stimulation methods	Electrical currents or chemicals may be used to stimulate specific points in the brain, and the effects on the animal's behavior studied.
Recording methods	Microelectrodes inserted into the brain allow the activity of neurons to be recorded during particular behavior. Electroencephalography uses electrodes on the scalp to record the coordinated activity of neurons.
Imaging	Computerized axial tomography scans the living brain with X-rays, providing a detailed picture of brain lesions. Positron emission tomography and functional magnetic resonance imaging produce pictures of brain activity.
Chemical approaches	Drugs that interfere with neurotransmitter function, or which mimic it, are used to study the neural basis of behavior. Pharmacological approaches also help to establish the influence of hormones by using drugs that interfere with or mimic aspects of hormone function. Anterograde labeling is used to show the efferent pathways, and retrograde labeling traces afferent pathways.
Genetic studies	The extent to which a particular behavior is determined by genetic factors can be assessed by family studies. Monozygotic twins have an identical genotype, dizygotic twins have 50% of their genes in common. Since environmental similarities usually vary in a parallel manner, some studies have compared monozygotic twins reared together with those adopted separately and reared apart. Selective breeding produces strains that differ in behavioral characteristics.
Related topic	Psychology and biology (A1)

Lesion studies One obvious way to find out what a part of the body does is to damage or remove that part and watch what happens. This approach has a long and fruitful history in physiology in general. In physiological psychology we are mostly concerned with studying damage to parts of the brain: **brain lesions**. Brain lesions may be produced by injury or disease, may be experimentally produced, or may result from therapeutic surgery. The aim is to see how an animal or person's behavior changes after an area has been lesioned.

We will see in Topic Q1 how the study of lesions resulting from disease has helped in the understanding of language function. Lesions in particular areas of the cerebral cortex cause distinctive changes in language abilities. However, we will also see a major disadvantage of relying on such information. Lesions resulting from accidental injury or disease processes are rarely limited to specific brain loci. This means that any one patient is likely to exhibit symptoms resulting from damage to several brain areas that might have distinct functions. A change in behavior might result from an unobserved lesion away from the focus of the main damage. This can lead to wrong conclusions about how particular psychological functions are organized in the brain.

The alternative is to produce lesions intentionally, which permits much smaller and more localized lesions to be studied. For obvious reasons, this is mostly done in animals. However, surgically-produced lesions have been investigated in human patients. For example, the different functions of the two cerebral hemispheres have been studied in people who have had the two hemispheres separated to relieve epilepsy (see Topic C3). Lesions may be produced by cutting, by high frequency electric current applied at the tip of an electrode, or by means of chemicals introduced into the brain through a tiny tube, a **cannula**. Chemical means may be used to selectively kill particular types of tissue. It is also possible to produce temporary lesions. An example of this in humans is the **Wada test**. A fast-acting barbiturate is injected into one or other carotid artery, temporarily anesthetizing the cerebral hemisphere on that side. The investigator can then examine performance or behavior when only one or other of the hemispheres is acting.

To assist surgeons and experimenters to produce lesions in precisely defined parts of the brain a **stereotaxic apparatus** is used. This fixes the head in position, and allows precise measurement to be made of the position of the electrode, wire, or cannula before the lesion is made. The exact location of lesions can usually only be described *post mortem*, using standard **histological** methods, for example microscopic examination of cross-sections of brain tissue. Until quite recently this hampered lesion studies in humans, as *post mortem* examination was often only possible long after the observations were made, or was not possible at all.

A general problem with lesion studies is in the interpretation of the results. As we will see in Topic M2, early investigators of the brain mechanisms of hunger claimed, using lesion methods, to have discovered 'feeding centers' in the brain. Later investigations showed that these sites are more generally involved in motivated behavior. It has also happened that the loss of a function has followed destruction of a pathway joining brain centers, and the pathway has been wrongly described as a control center. Further, we need to be careful that the brain area destroyed controls the target behavior, and not some other behavior essential for the occurrence of the target behavior. To use an extreme example, suppose we taught an animal to make a choice based on a visual cue. If we then destroyed the primary visual cortex (see Topic G1) and showed that the animal could no longer make the choice, we should not conclude that the learning had taken place in the visual cortex. The animal would fail simply because it was blind.

Finally, performance of a particular type of behavior might involve several components. Performance can be disrupted by lesions that interfere with any one component, leaving other components of the behavior unaffected. This **double dissociation** is shown, for example, in reading, when lesions that affect spatial abilities can occur independently of lesions that affect semantic ones (see Topic Q2).

Stimulation methods

A less destructive way of studying the functions of brain sites is to use **stimulation studies**. Weak electrical currents or chemicals are used to stimulate specific points in the brain, often using a stereotaxic apparatus, and the effects on the animal's behavior are studied. We will see examples of this in Topic P2, where we see how rats learned to stimulate particular locations in their own brains, leading to the identification of reward circuits. The method has also been used with conscious humans. Electrical stimulation of the exposed brain during neurosurgery has provided important information about different areas of the cortex (see Topic Q2).

Recording methods

The converse of stimulation methods are **recording methods**. These allow us to find out which parts of the brain are active during particular behaviors. Stereotaxic apparatus is again used to insert very fine **microelectrodes** into the brain. From these the activity of neurons may be recorded while the animal performs particular behaviors. It is possible to refine this method so that single neurons are recorded: **single unit recording**. Another approach is to place electrodes on the scalp, recording the coordinated activity of large numbers of neurons in the underlying brain regions. This technique, **electroencephalography** (EEG) was introduced in 1929 by Hans Berger. The EEG has been used to study perceptual processes and sleep (Topic K1), amongst other things, in animals and humans.

Imaging

In the last 30 years a breakthrough in the study of brain function in intact animals and people has resulted from the introduction of a number of imaging techniques. **Computerized axial tomography** (**CAT scan**) scans the living brain with X-rays, and allows a detailed picture of brain lesions to be produced. This has done away with the need to wait for *post mortem* examinations. However, CAT scans show static structures, rather than brain activity in relation to function, which is often of greater interest in physiological psychology.

Two newer techniques do reveal changes in the brain related to function. Blood flow is increased in active brain areas. Therefore, if water containing radioactive oxygen is injected, more of it reaches active brain areas. **Positron emission tomography** (**PET scan**) produces an image of the effects of this radiation in the brain. The second technique, **functional magnetic resonance imaging** (**fMRI**), assesses the different excitement of hydrogen nuclei in different molecules, produced by a magnetic field. When a tissue is active, less hydrogen is contained in **deoxyhemoglobin** molecules, changing the relative excitement of the nuclei. When the brain is scanned and imaged quickly enough, this gives detailed information about the activity of particular brain regions. We will see numerous examples of how the use of these methods has revealed which parts of the brain are active in a wide range of psychological processes.

Chemical approaches

The main purpose of the methods discussed in preceding paragraphs is to locate where particular functions are controlled or represented in the brain. In many cases it is useful to find out what chemical substances are involved in psychological processes. As we will see in Topic B3, chemical **neurotransmitters** pass messages from one brain cell to another. Drugs that interfere with various aspects of this, or which mimic neurotransmitter action, are used to study the neural basis of various states and behaviors. We will see, for example, how clues to the nature of certain psychological disorders were obtained by studying how therapeutic drugs act (see Topics R1 and R2). Such pharmacological approaches also help to establish the influence of hormones on behavior (see Section E). Drugs that interfere with or mimic aspects of hormone function have shown, for example, the mechanisms underlying the menstrual cycle (Topic E6).

A different use of chemical methods is to trace neural pathways connecting different brain centers. The use of chemical stains and histological examination allows some progress to be made in this. However, better results have been obtained by **anterograde labeling**. A number of substances, one of which is called **PHA-L**, are absorbed by neurons and passed into the axon towards the terminal boutons (see Topic B1). In this technique, a small amount of PHA-L is injected into a neuron, the animal is killed, and its brain is stained using a reagent that makes the PHA-L visible. This is examined microscopically, and shows efferent pathways. A similar technique, **retrograde labeling**, uses substances (such as **fluorogold**) that are absorbed by terminal boutons and passed back to the cell body. This allows the tracing of afferent (sensory) pathways. As an example, we will see in Topic N3 how these methods have shown the pathways involved in the control of sexual behavior.

Another chemical method that lets researchers trace neural circuits is based on the fact that genes in active neurons cause characteristic proteins, such as **Fos**, to be produced in the terminal boutons. Measuring the amount of Fos produced during a behavior or manipulation indicates which neurons are active. We will see an example of this when we look at the neural control of sexual behavior (Topic N3).

Genetic studies   There are many ways of trying to establish the extent to which a particular behavior is determined by genetic factors. In humans, most studies look at how similar people are who have different degrees of family relationship. **Monozygotic** (identical) **twins**, who have an identical genotype, should be the most similar. **Dizygotic** (fraternal) **twins** have 50% of their genes in common, just as any other siblings. The proportion of shared genes decreases as relationships become more distant. Of course, environmental similarities usually vary in a parallel manner, making it difficult to disentangle the two influences. To help separate environmental and genetic effects, some studies have compared monozygotic twins reared together with those adopted separately and reared apart. These techniques have been particularly useful in estimating the genetic influence on mental illness (see Section R). In other species, selective breeding produces strains that differ in behavioral characteristics, demonstrating a genetic basis for those characteristics. More directly, **molecular biology** examines the actual genetic differences between animals and people with different characteristics. We will see many examples of all of these approaches in the later sections of this book.

B1 THE COMPONENTS OF THE NERVOUS SYSTEM

Key Notes

Cell body	The cell body of a neuron contains all of the components that allow the cell to function (e.g. mitochondria and ribosomes). It also contains the nucleus. The cell body of a neuron is specialized for its particular functions.
Dendrites	These are the input side of the neuron. They often look like the branches of a tree in winter. They have specialized areas for receiving chemicals released by other neurons.
Axon	There is usually only one axon emerging from the cell body. This is the output side of the neuron. The axon can branch in order to supply its output to more than one place.
Axon hillock	This is where the output signal gets generated. If the conditions at this point of the neuron are not right, then the neuron will not fire an impulse along the axon.
Terminal bouton	This is the thickening of the neuron at its distal (far) end. This thickening houses the synaptic vesicles that store neurotransmitter proteins.
Synapse	This is the region at which communication occurs between the neuron and whatever it connects to. This might be another neuron or a different type of body structure, such as a muscle or a gland.
Neuronal membrane	The walls of the neuron are called the neuronal membrane. It is made up of two layers of fat but has numerous different structures embedded into it. These might be channels for the movement of chemicals or specialized proteins such as receptors.
Myelin sheath	A protective layer surrounds some axons. This is called a myelin sheath.
Related topics	Transmission of the nerve impulse (B2) Synaptic transmission (B3)

Cell body

The whole **neuron** is illustrated in *Fig. 1*. The cell body (or **soma**) of a neuron contains a number of different **organelles** that all work towards the maintenance of the cell. Some of the more important organelles within the cell body are the **nucleus**, which houses the cell's DNA, the **mitochondria**, which provide the cell with energy, and the **ribosomes**, on which proteins are produced.

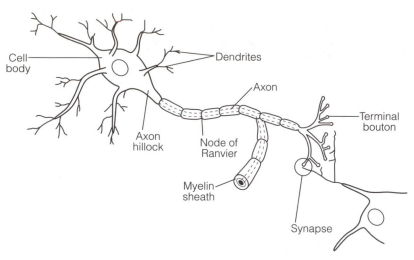

Fig. 1.   The features of a typical neuron.

Whilst the cell body of a neuron is not very different to the cell body of other cells, neurons are specialized to produce very large amounts of protein. For this reason, there are large numbers of ribosomes which are packed densely together on the **endoplasmic reticulum**. These appear as distinct structures called **Nissl bodies**.

Dendrites

Every neuron has an input and an output. The **dendrites** are the input side of the neuron. In many neurons they appear as a system of branches, although how dense and extensive this branching is will depend on what type of neuron it is. These branches arise from a number of different points on the cell body. If one looks at the branches through an electron microscope, each branch has numerous regions which are thicker than the rest of the branch. These regions are called the **postsynaptic thickening** and it is here that the postsynaptic receptors are located (more details follow in B3). The branching dendrites allow the neuron to be influenced by the impulses being sent from many other neurons and this is the process of **convergence**. So, for example, within the visual system, object recognition is the result of the convergence of separately processed information about shape, depth, size and motion.

Axon

less branches than dendrites

The **axon** is the output side of the neuron. Unlike the dendrites, there is usually only one axon emanating from each cell body. This single axon can also branch, but there are usually far fewer branches to the axon of a neuron than to its dendrites. Indeed, some axons do not branch at all. Axons can extend for just a few microns (millionths of a meter) or for several feet (as in the case of a sensory axon that starts in your big toe and ends up in the brain). It is the axon that carries the nerve impulse to whatever that neuron innervates (e.g. a muscle, a gland, or even another neuron). Where an axon does branch, the information will be transmitted to more than one place. This allows for **divergence** of information and this is complementary to the convergence explained for dendrites.

Axon hillock

The **axon hillock** is located where the axon leaves the cell body. It is a special place on the neuron because this is where the neuronal impulse, which will be carried along the axon, is generated. We shall see later in this section (B2) that particular electrochemical conditions must exist at the axon hillock in order for a neuron to fire.

Terminal bouton

At the end of the neuron that is farthest from the cell body (the distal end) the axon swells to form a **terminal bouton**. Within this bouton are specialized sacs, called **vesicles**, for storing **neurotransmitter** molecules (see Topic B3). It is here that the electrochemical signal that travels along the axon is converted into a chemical signal that can be transmitted from the neuron to the cell(s) it connects to. The chemical signaling is achieved by the release of the stored neurotransmitter molecules from the end of the terminal bouton. If an axon divides to form multiple branches, each branch ends in a terminal bouton.

Synapse

A **synapse** consists of the terminal bouton of one neuron, the receptor area of the next cell (another neuron, or a muscle or gland), and the space in between them, called the **synaptic cleft**. The three areas that make up a synapse are the **presynaptic terminal**, the **synaptic cleft**, and the **postsynaptic membrane**. The synapse is the region where a neuron communicates with whatever it connects to. The communication across most synapses is achieved by way of the release of chemicals called neurotransmitters (see Topic B3). There are also electrical synapses where the communication between cells is electrical rather than chemical. The membranes of the two cells are close enough together that the ions can diffuse directly from one cell to the next. The place at which the two membranes meet is called a **gap junction**.

Neuronal membrane

The neuronal membrane itself is made of a bilipid layer (i.e. two layers of fat molecules). A number of different protein structures are embedded into this fatty membrane. Some protein molecules act as receptors that register the presence of one or more types of neurotransmitter molecule. Others provide channels through the membrane along which ionized atoms and some large molecules can pass from one side of the membrane to the other. Many of these proteins can change their shape to open or close the channel depending on the prevailing electrochemical conditions.

Myelin sheath

Most, but not all, neurons have a protective layer around the axon. This layer, the **myelin sheath**, is made up of **Schwann cells** in the peripheral nervous system and **oligodendroglia** in the central nervous system. The protective cell surrounds the axon like an onion skin. It not only serves as a protective layer but also stops the membrane leaking ions across it. However, given that ions must cross the membrane sometimes (see Topic B2), there are breaks in the sheath every so often along the axon's length. These breaks are called **nodes of Ranvier**, and in the mammalian nervous system there is a node of Ranvier at least every 2 mm.

B2 TRANSMISSION OF THE NERVE IMPULSE

Key Notes

Resting membrane potential	This is the voltage across the neuronal membrane when the neuron is at rest. Its value is usually around –70 mV, with the inside being negative with respect to the outside. The voltage is generated by the concentration and electrical gradients of ions on either side of the membrane.
Action potential generation	An action potential is generated at the axon hillock. It occurs when the voltage across the neuronal membrane is sufficiently depolarized to reach a threshold level. This threshold is usually around –55 mV (inside negative).
Action potential	An action potential can be measured as the changes in voltages across the neuronal membrane that occur when the neuron is active (i.e. when it is transmitting an impulse). The action potential is caused by the movement of ions across the membrane and this ion movement has a characteristic time course.
Conduction along the axon	An action potential is conducted along an axon by a succession of depolarizations along its length. In a myelinated axon the action potential jumps from one node of Ranvier to the next (a process called saltatory conduction). Once an action potential has passed along the axon, that axon can no longer be stimulated (absolute refractoriness) or can only be stimulated by a stronger depolarization at the axon hillock (relative refractoriness).
Related topics	The components of the nervous system (B1) Synaptic transmission (B3)

Resting membrane potential

Just as with any electrical circuit, we can measure voltages from neurons. If we measure the voltage across the neuronal membrane, even when the neuron is at rest, there is a voltage of –70 mV. This voltage represents the fact that the inside of the neuron has a small negative charge relative to the outside of the neuron. In order to consider how this voltage difference is generated, we need to consider what an **ion** is. When we place a molecule (e.g. sodium chloride, which has the symbol NaCl and is better known as common salt) into water, the molecule breaks apart into charged particles called ions. In the case of NaCl, the molecule breaks apart into a positively charged sodium ion (Na^+) and a negatively charged chloride ion (Cl^-).

The **resting membrane potential** depends upon the fact that the neuronal membrane is more permeable to some ions than to others. For example, big, negatively charged protein ions are trapped inside the neuron. However, smaller potassium (K⁺) ions can be actively or passively transported across the membrane. In the resting state, the distribution of ions on either side of the membrane sets up two forces that act on these ions to either drive them across the membrane or keep them where they are. One such force is the **chemical gradient** (*Fig. 1a*). This tries to move ions from areas of high concentration into areas of low concentration. The other force is the **electrical gradient** (*Fig. 1b*). This tries to move ions towards an area that has the opposite charge to the charge on the ion. This is similar to the magnetic force, in that opposites attract and likes repel.

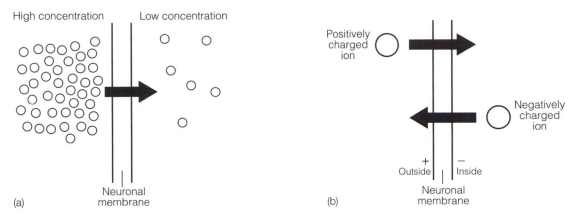

Fig. 1.   (a) A chemical gradient. (b) An electrical gradient.

To examine the effect of these forces on the resting membrane potential, we need to consider two particular ions and their relative concentrations either side of the membrane. Sodium ions (Na⁺) are in higher concentration outside of the neuron, whereas potassium ions (K⁺) are in higher concentration inside the neuron. Hence the chemical concentration gradient tries to force Na⁺ ions into the neuron and K⁺ ions out of the neuron. The story for the electrical gradient is different. As the inside of the neuron is negatively charged, the electrical gradient tries to drive Na⁺ ions into the neuron but tries to prevent the K⁺ ions from escaping. The net effect (*Fig. 2*) is that both forces try to drive Na⁺ ions into the neuron but the two forces oppose each other as far as K⁺ ions are concerned.

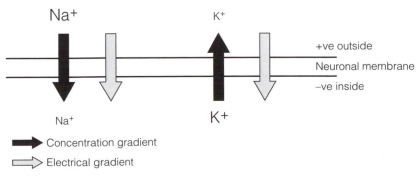

Fig. 2.   The neuronal membrane effects of the chemical and electrical gradients on sodium and potassium ions.

What prevents mass ion movement across the membrane in the resting state? The answer is that most of the channels for ion movement are active channels and these channels are **voltage-gated**. This means that the channels are only open when certain voltage conditions occur. In the resting state, the gates are shut. There is some ion movement but this is referred to as *leakage*. This leakage is reversed by a **sodium–potassium pump** that pumps 3 Na⁺ molecules out of the neuron in return for pumping 2 K⁺ ions back in.

Action potential generation

When a neuron is stimulated, for example as a result of another neuron having fired, each element of stimulation results in only a very small and local change in the neuron's voltage (considered later in the chapter for synaptic transmission). If the stimulation is *excitatory*, the voltage will shift very slightly from –70 mV towards zero (the membrane is said to be **depolarized**). This local event is called an **excitatory postsynaptic potential** (**EPSP**) (*Fig. 3a*). If the stimulation is *inhibitory*, the voltage will shift very slightly from –70 mV towards –80 mV (the membrane is said to be **hyperpolarized**). This local event is called an **inhibitory postsynaptic potential** (**IPSP**) (*Fig. 3b*). Whether or not our postsynaptic neuron fires an action potential depends upon the voltage change reaching a **threshold value** (often around –55 mV). This threshold value must be reached at the axon hillock (see Topic B1, *Fig. 1*), the point at which the axon leaves the cell body. The voltage at the axon hillock at any given time is the net effect of all the local excitatory and inhibitory voltage changes. Hence, in order for a neuron to fire, there have to be either many local excitatory stimulations (compared to inhibitory stimulations) at the same time or a rapid succession of a smaller number of excitatory stimulations (compared to inhibitory stimulations) at the same location. The former is called **spatial summation** (*Fig. 4a*) and the latter is called **temporal summation** (*Fig. 4b*). Finally, if the threshold value is reached at the axon hillock, nothing can then prevent the neuron from firing

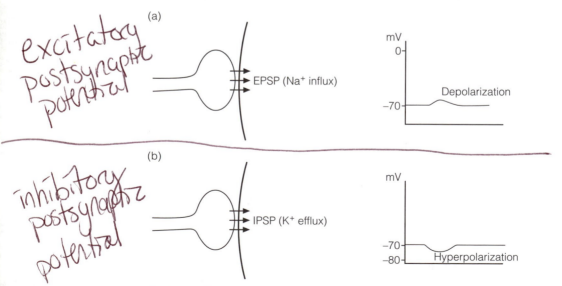

Fig. 3.   (a) The effect and time course on membrane voltage of sodium movement across the membrane. (b) The effect and time course on membrane voltage of potassium movement across the membrane.

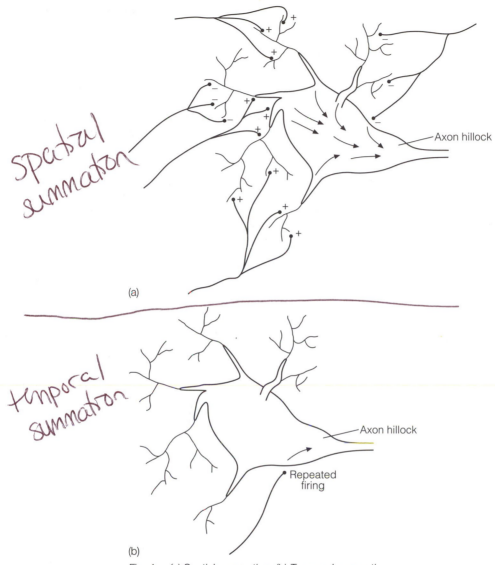

spatial summation

temporal summation

(a)

Axon hillock

(b)

Axon hillock

Repeated firing

Fig. 4.   (a) Spatial summation. (b) Temporal summation.

an action potential. In addition, when the neuron fires, the size of the action potential for a given neuron is always the same. For these reasons, the action potential is referred to as an **all-or-nothing** event.

Action potential   We can examine an action potential in two different ways. Here, we examine how the action potential causes ions to move across the membrane at a given point on the neuronal membrane. Under the next heading, we examine how the action potential travels along the axon.

Remember the ionic scenario that we had when we considered the resting membrane potential: the outside of the neuron is rich in sodium (Na^+ ions) and the inside is rich in potassium (K^+) ions. Remember, also, that the voltage-gated channels are shut. When an action potential reaches a point on the membrane, the change in voltage activates ion channels in a particular sequence (see *Fig. 5*). The first to become activated are the sodium channels. They open slightly, at first, and sodium ions pass through the gates and into the neuron. This causes a further depolarization and the gates open a little further. As the depolarizing current here is large enough (*suprathreshold*), the gates will open fully and sodium ions will flood into the neuron. This creates the *rising phase* of the action potential. After around 0.5 msec the potassium channels open and these allow K⁺ ions to flood out of the neuron. About 1 msec after the sodium channels have opened, the gate slams shut. The voltage across the membrane at this point is around +40 mV. Now only the potassium gates remain open and so the voltage starts to return towards its resting value. However, the potassium gates do not shut until the voltage has gone beyond its resting value to –80 mV. The sodium–potassium pump works quickly to restore the balance and the resting membrane potential is re-established. This whole sequence gives the characteristic action potential curve shown in *Fig. 5*.

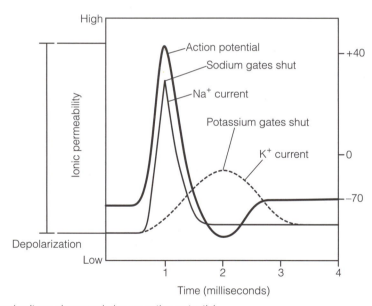

Fig. 5.   *Ionic and voltage changes during an action potential.*

Conduction along the axon

We have seen how an action potential is characterized at a single point along the axon's length. How does this translate into an action potential that flows along the axon from one end to the other?

If we consider a point on the membrane where depolarization has occurred, we will find the concentration of sodium ions is higher inside the neuron here than it is elsewhere. Due to the concentration gradient, these sodium ions will be pulled to further along the inside of the neuron where the concentration is lower. This will cause the electrical potential at this new point to become slightly depolarized. This in turn will open slightly the voltage-gated channels, letting sodium ions start to enter from outside of the neuron. Very soon, the neuron at this point will be depolarized beyond the threshold value and the action potential will have moved along the axon.

Why does the action potential not move in both directions? At the time when the sodium ions have started to depolarize the next part of the neuron, there is still a high level of depolarization a bit further back. When sodium ions a little further down enter the axon they are repelled away from this region by both the electrical and concentration gradients (*Fig. 6*). A little further back still, the membrane is in a state where it is hyperpolarized. This sets up three different regions along the activated neuron. In one place the membrane is at rest, in a second place the action potential is happening, and in a third place the membrane is hyperpolarized. The latter two places have what is referred to as **refractoriness**. Where the neuron is activated, it cannot be stimulated again at all. The membrane here is said to be **absolutely refractory**. Where the membrane is hyperpolarized, it can be stimulated again but only if the stimulus

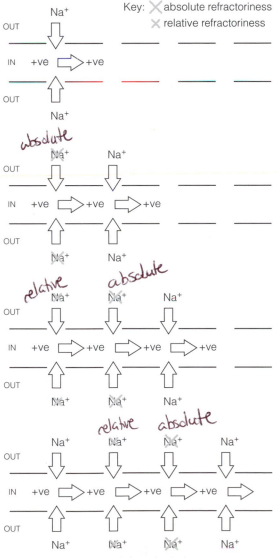

Fig. 6. Absolute and relative refractoriness.

is larger than one that would normally be required when the neuron was at rest. The membrane here is said to be **relatively refractory** (*Fig. 6*).

Most axons are protected by the myelin sheath (see Topic B1). This prevents ions from leaking in or out of the neuron. In order that ions can pass through where required, the sheath has breaks in it every so often at the nodes of Ranvier. In axons with a sheath, the action potential jumps from node to node and this type of conduction is called **saltatory conduction**. Saltatory conduction is faster than conduction along an unmyelinated axon.

B3 SYNAPTIC TRANSMISSION

Key Notes

The synapse	The synapse comprises the presynaptic region, the synaptic cleft and the postsynaptic region. It is here that chemical transmission takes place between the neuron and the postsynaptic cell.
Neurotransmitter substances	Neurotransmitter substances are molecules that are released from the end of the terminal bouton when the conditions are right. There are different types of neurotransmitter substance and the decision as to whether a molecule is or is not one is made according to a set of criteria. Their purpose is to facilitate communication between the neuron and whatever lies beyond it.
Release of neurotransmitter substance	Neurotransmitter molecules are made in the cell body but are stored in the terminal bouton inside vesicles. The vesicles are released from their fixings in the terminal bouton by calcium influx. They then migrate to the cell membrane and release the neurotransmitter molecules by a process of exocytosis.
Action at receptors	The neurotransmitter molecules travel across the synaptic gap and attach themselves to specific receptors on the postsynaptic membrane. Neurotransmitter molecules can act as first or second messengers. The former give rise to the opening of ion channels causing localized depolarization or hyperpolarization.
Inactivation of neurotransmitter substances	After detachment of the neurotransmitter from its receptor occurs, there is enzymatic degradation or reuptake of the neurotransmitter molecules to prevent their re-activating the receptors. There is some reuse of some of the products of neurotransmitter degradation.
Related topics	The components of the nervous system (B1)　　Transmission of the nerve impulse (B2)

The synapse

The region called the **synapse** is made up of the presynaptic terminal, the post-synaptic membrane region, and the small space between them known as the **synaptic cleft** (*Fig. 1*). The synaptic cleft is an extremely small gap (about 20 nm – a nanometre, nm, is one billionth of a meter). The postsynaptic membrane contains specialized receptors to which the correct neurotransmitter substance can attach.

Neurotransmitter substances

When an action potential reaches the end of the axon, the message it carries is converted from an electrical signal to a chemical signal in order for the message to be transmitted to the next cell. The chemical substances that carry these messages are called neurotransmitter substances. Given that many molecules and ions come and go from the neuron, there is a set of criteria for judging

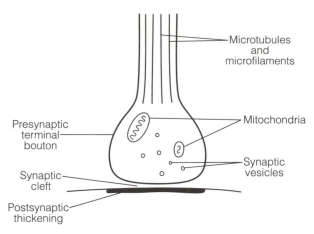

Fig. 1.   An expanded view of a synapse.

whether or not a molecule satisfies the conditions for being called a **neurotransmitter substance**. These are:

● the substance is synthesized in the neuron's cell body;
● it is stored in the **synaptic vesicles**;
● it is released from the neuron by a mechanism which is calcium-dependent;
● a synthetic neurotransmitter applied exogenously must precisely mimic the actions of the true neurotransmitter;
● there must be a mechanism for rapid termination of the action of the released neurotransmitter.

Well-known neurotransmitter substances that do fit these criteria are acetylcholine, norepinephrine (sometime called noradrenaline), dopamine, serotonin, glutamate, and gamma-aminobutyric acid (GABA). However, there are molecules that fulfill most but not all of these criteria and are probably neurotransmitter substances. These are called **putative neurotransmitters**. There are numerous substances that fit this label but some of the more interesting ones are nitric oxide, substance P, and the endorphins.

Release of neurotransmitter substance

Neurotransmitter substances are made in the cell body of the neuron but they are stored in the terminal bouton inside sacs called **synaptic vesicles**. These sacs are fixed to internal structures inside the bouton when the neuron is at rest. When an action potential reaches the terminal bouton, calcium channels open and calcium ions (Ca^{2+}) flood into the bouton. These ions release the vesicles from their fixings and the vesicles migrate to the end of the bouton. Once there, the vesicles fuse with the neuronal membrane to release the contents into the synaptic cleft (*Fig. 2*). This process is called **exocytosis**.

Action at receptors

Once the neurotransmitter molecules are released into the synaptic cleft, they diffuse across the space and some land on receptors located in the postsynaptic membrane. The receptor and the neurotransmitter molecule can be thought of as a lock and key mechanism: if the correct molecule attaches to the correct receptor a postsynaptic event will occur. There are two types of response that can be made by the postsynaptic receptor mechanism. One is called the **first messenger** system and this leads to ion channels being opened to let ions flow

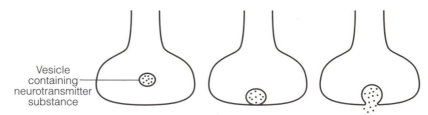

Vesicle containing neurotransmitter substance

Fig. 2.   The principle of exocytosis.

in or out of the postsynaptic region. If the postsynaptic cell is another neuron then this would be the mechanism that generates the excitatory and inhibitory postsynaptic potentials (EPSPs and IPSPs) referred to in Topic B2. The other type of postsynaptic receptor mechanism is the **second messenger** system. This is much more complex and involves a cascade of internal events that often lead to changes in the morphology or functioning of the postsynaptic cell. For example, in some animals, sensitization (a form of simple learning) is believed to be the result of the closure of a potassium channel by the action of a second messenger called cAMP.

In first messenger systems, whether or not the net effect of an input is excitatory or inhibitory can be related to the particular neurotransmitter that acts at that synapse. For example, dopamine is predominantly an excitatory neurotransmitter so it tends to produce EPSPs through its action at the receptor sites. In contrast, GABA tends to be inhibitory and so tends to produce IPSPs. However, it should be remembered that the interconnections between neurons are complex, and an overall inhibition could just as easily be the result of exciting an inhibitory pathway as inhibiting an excitatory one.

Inactivation of neurotransmitter substances

The process of communication we have been describing is designed to pass discrete messages between a neuron and the next cell. If the neurotransmitter substance stayed attached or re-attached to its receptor for any length of time then this message might become confused with the next one that comes along. Hence there needs to be a mechanism for the rapid inactivation of the neurotransmitter molecules associated with a single action potential. In fact, two slightly different mechanisms exist.

Once the neurotransmitter molecule has become detached from the receptor it is either deactivated in the synaptic cleft by specific enzyme molecules or it is taken back up into the terminal bouton to be deactivated there. Deactivation by an enzyme often leaves a waste product and a product that can be re-used to make more neurotransmitter molecules. For example, acetylcholine is broken down into acetic acid and choline within the synaptic cleft by an enzyme called acetylcholinesterase. The choline can be taken back up into the presynaptic terminal and re-used to make more acetylcholine.

C1 THE ANATOMY OF THE CENTRAL NERVOUS SYSTEM

Key Notes

Definition, planes and sections of the central nervous system	The central nervous system (CNS) consists of the brain and spinal cord. Within the brain, cell bodies are described as gray matter. The 3-dimensional planes of the brain are: top to toe (anterior–posterior); from back to front (dorsal–ventral); from the mid-line to the sides (medial–lateral). Within the brain, rostral means further from the spinal cord, and caudal means nearer the spinal cord.
The major divisions of the brain	The brain can be divided up for classification in many different ways. We use the division into hindbrain, midbrain, and forebrain. The capillaries in the brain impose a blood–brain barrier preventing many large molecules reaching the brain.
The hindbrain	The hindbrain consists of the medulla oblongata, the pons and the cerebellum. Between them, these structures control the vital functions of the organism.
The midbrain	The midbrain consists of the tectum and the tegmentum. The tectum contains the colliculi and the tegmentum includes structures like the periaqueductal gray matter, the substantia nigra and the red nucleus.
The forebrain	The forebrain contains many individual components spanning two hemispheres connected by the corpus callosum. The surface region is called the cerebral cortex. Important subcortical structures include the thalamus, whose main role is as a relay station and processing center for information coming to the cortex; the hypothalamus which is crucial for our internal regulation; the basal ganglia; the limbic system; the hippocampus; and the amygdala.
Lobes of the brain	The forebrain can be conveniently divided into four lobes, the frontal lobe, the parietal lobe, the temporal lobes and the occipital lobe.
The cerebral ventricles	As well as brain tissue, the cranium contains a series of fluid filled canals called the ventricles. These carry cerebrospinal fluid and this fluid serves as a support system for the brain.
Related topics	Localization of cortical function (C2) Hemispheric lateralization (C3)

Definition, planes and sections of the central nervous system

The central nervous system (CNS) consists of the brain and the spinal cord. All other nerve cells are part of the peripheral nervous system. The constituent components of the brain are collections of cell bodies that appear gray when looked at under the microscope (the gray matter) and fiber tracts that appear white under the microscope (white matter). Hence structures in the brain are all discrete regions of gray matter.

In order to understand the naming of brain structures we need to consider the directional planes used to describe it. *Fig. 1* shows the three planes. Top to toe is the plane described as **anterior–posterior**. From back to front is the plane described as **dorsal–ventral**. Finally, from the mid-line to the sides is the plane described as **medial–lateral**. The terms anterior, posterior etc. are used to describe the locations of structures in the CNS. If we consider the torso, these planes are easy to identify. In humans the brain is folded over at its 'top' end, as shown in *Fig. 2*. This means that the brain tissue lying immediately under the forehead is not really at the front (ventral) but at the top (anterior). This strange state of affairs need not cause any problem but it may mean that the relative labels of adjacent parts of the brain are counter-intuitive.

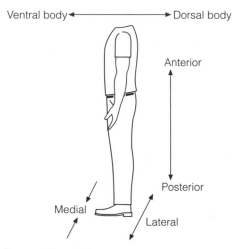

Fig. 1.   Planes of anatomical directions in the human body.

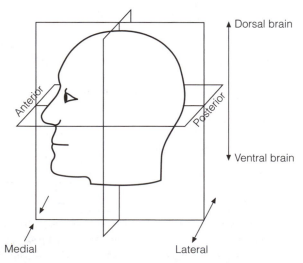

Fig. 2.   Planes of anatomical directions in the human brain.

In order to see the positioning of structures that lie under the surface, we have to look at cut slices of the brain. Given that there are three dimensions, there are three planes in which a slice can be cut. If we make our cut so that it goes from front to back down the middle of the nose, the cut end would look like that in *Fig. 3*. This is known as a **sagittal section**. If we make the cut side to side going down through the ears, the cut end would look like that in *Fig. 4*. This is known as a **coronal section**. Finally, if we make the cut as though we were taking the top off a boiled egg, the cut end would look like that in *Fig. 5*. This is known as a **horizontal section**.

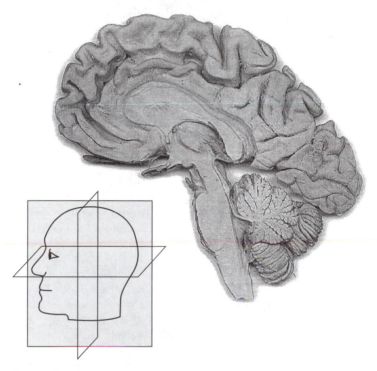

Fig. 3.   A sagittal section.

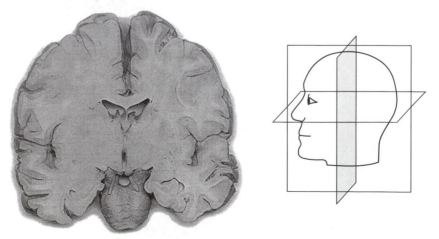

Fig. 4.   A coronal section.

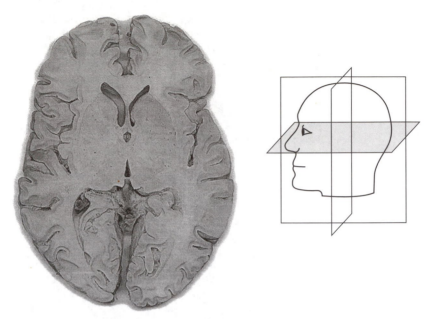

Fig. 5.   A horizontal section.

We will use two additional terms to describe positions of structures relative to one another within the CNS. **Rostral** (literally 'towards the beak') is used to mean further from the spinal cord. **Caudal** ('towards the tail') means nearer the spinal cord.

The major divisions of the brain

The brain can be considered to be divided into its main areas in several ways. Most accurately, the brain can be divided into five regions, the **myelencephalon**, the **metencephalon**, the **mesencephalon**, the **diencephalon**, and the **telencephalon** (*Fig. 6*). The last of these are the **cerebral hemispheres**. The other four are sometimes collectively known as the **brain stem**.

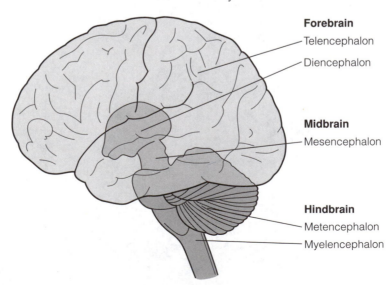

Forebrain
Telencephalon
Diencephalon

Midbrain
Mesencephalon

Hindbrain
Metencephalon
Myelencephalon

Fig. 6.   Major structures of the human brain.

For our purposes it is simpler to use a classification that has three parts. The myelencephalon and metencephalon are known together as the hindbrain. The mesencephalon is the midbrain and the diencephalon and telencephalon together are the forebrain. Each of these regions has a large number of structures. In the next three parts, only the major structures will be discussed.

The blood supply to the brain has an important feature different from that to other tissues. In order to reduce the number of potentially toxic chemicals that can enter the brain there is a **blood–brain barrier**. This is a feature of the walls of the capillaries in the brain. Their construction is such that they do not easily let large molecules through. It is for this reason, for example, that Parkinson's disease patients have to be given L-Dopa (a dopamine precursor) rather than dopamine as the dopamine will not pass through the blood–brain barrier.

The hindbrain

The hindbrain is the first part of the brain that you come to once you leave the spinal cord. It consists of three separate structures, the medulla oblongata, the pons and the cerebellum (*Fig. 7*). Nuclei within the medulla oblongata carry out a number of vital functions such as the regulation of heart rate and breathing. The word pons is Latin for bridge. As this name suggests, the pons is one of the regions of the brain where nerve fibers cross the midline from one side of the body to the other. It is here, for example, where sensory signals from the left side of the body cross to go to the right somatosensory cortex. The cerebellum is involved in movement-related functions. It is the seat of at least some motor programs that allow us to carry out well-rehearsed movements automatically. It is also the region of the brain that interprets balance information.

Given that the hindbrain is, in an evolutionary sense, the oldest part of the brain, it is hardly surprising that this region is involved in all of the functions that are vital to an organism's survival. Without control of the vital organs and without simple capabilities in sensation and locomotion, no organism could survive.

The midbrain

The midbrain has a dorsal region called the **tectum** and a ventral region called the **tegmentum**. The tectum houses the **inferior colliculi** which have a role in audition (see Topic H1) and the **superior colliculi** which have a role in vision (see Topic G1). In lower vertebrates the whole of the tectum is devoted to the visual modality.

The tegmentum includes the **periaqueductal gray matter** (**PAG**) which has an important role in analgesia. More specifically, the PAG contains opiate receptors that respond to our endogenous opiate neurotransmitters (see Topic F3). Also in the tegmentum is the **substantia nigra**. This nucleus plays an important role in motor control. It is the substantia nigra that is damaged in Parkinson's disease. The **red nucleus** is also a motor control structure located in the tegmentum. It plays an important role in motor output and is a major output destination from the cerebellum.

The forebrain

There are far too many structures in the forebrain for them all to be considered here. Some of the more important ones are illustrated in *Fig. 8*. The forebrain is made up of two hemispheres (left and right) that are joined together by a large fiber tract called the **corpus callosum**. The outermost layer of the brain is called the **cerebral cortex** or **neocortex**. This is of crucial importance to higher animals as it is where the most complex information processing takes place. However, this cortical layer is only about 6 mm thick in humans and so most of the cerebral hemispheres are made up of subcortical structures.

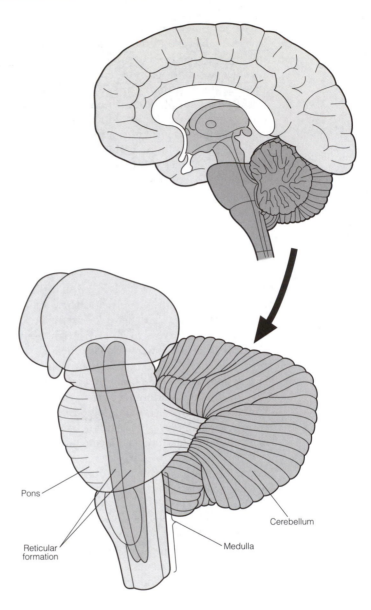

Fig. 7.   *Important structures in the hindbrain.*

One classical way to divide up the forebrain is into the diencephalon and the telencephalon. The diencephalon consists mainly of the **thalamus** and the **hypothalamus** (see *Fig. 8 and Fig. 9*). The thalamus is itself made up of numerous separate nuclei. Its main role is as a relay station and processing center for information coming to the cortex. The hypothalamus is crucial for our internal regulation (homeostasis; see Topic L1) and will feature strongly in later sections of the book.

The telencephalic structures (*see Fig. 9*) that lie outside of the cortex are best described here in terms of neural systems. A motor system, called the **basal**

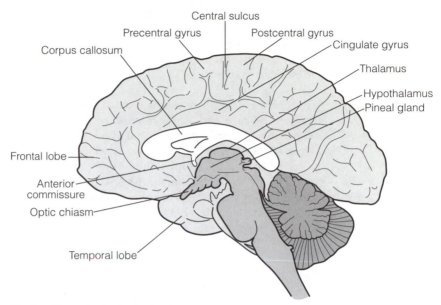

Fig. 8.   The major forebrain structures.

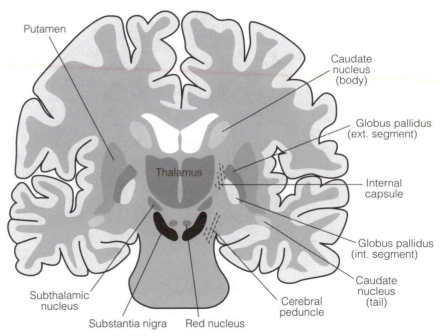

Fig. 9.   Major structures of the diencephalon and the telencephalon.

division of forebrain

ganglia, is made up of the **caudate nucleus**, the **putamen** and the **globus pallidus**. The caudate nucleus and putamen (the neostriatum) are innervated by the **substantia nigra**, and it is the cells projecting from the substantia nigra to the caudate nucleus that are damaged in Parkinson's disease.

A major system in the forebrain is the limbic system (see Topic O2, *Fig. 2*). This comprises its own cortex (the **limbic cortex**) that is distinct from the neocortex described above. An important part of this cortex is an area called the **cingulate gyrus**. Other structures that make up the limbic system are the hippocampus, the **amygdala**, the **mammillary bodies**, and a fiber tract called the **fornix**. This system is involved in emotion and memory and its structures will be referred to in Topics O2 and P3.

Lobes of the brain

Each cerebral hemisphere can be divided into four **lobes** (*Fig. 10*), composed of numerous folds, or **gyri**. At the front of the brain is the aptly named **frontal lobe**. This extends as far back as the **central gyrus**. Behind this, on the top of the brain is the **parietal lobe**, whilst behind this at the sides is the **temporal lobe**. At the very back of the brain is the **occipital lobe**. The frontal lobe has numerous functions, as we will see in later sections of this book. Where the frontal lobe ends in the region of the central gyrus is the primary motor area. A little further forward from this, but only on the left side, is an area called **Broca's area**. This is a region of the brain that is responsible for the production of speech (see Topic Q2).

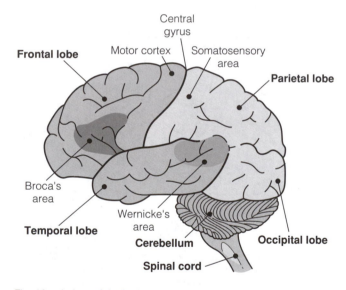

Fig. 10.   Lobes of the brain.

The parietal lobe contains much of what is referred to as the **association cortex**. It contains many of the regions of the brain where information is integrated after it has been processed for its initial perceptual qualities.

The temporal lobes are implicated in **hearing** and **memory** functions. Indeed, damage to the temporal lobes is associated with a form of amnesia. On the left side of the brain, the temporal lobe contains an area of the brain called **Wernicke's area** (see Topic Q2). This area is believed to play an important role in language comprehension.

The lobe at the very back of the brain is the occipital lobe. This region of the brain is exclusively devoted to vision. It contains a **striate region** that is the primary visual area and a **prestriate area** that is involved in the further processing of visual information.

The cerebral ventricles

Running through the center of the spinal cord and up into the brain is a fluid-filled region. This allows the brain and spinal cord to be supplied with nutrients and for waste products to be taken away. Within the brain this region comprises the **cerebral ventricles**. The pressure in the fluid filled area also allows for the brain's shape to be maintained. Indeed, one way of determining that brain damage has occurred is by recognizing that the ventricles are enlarged (i.e. they occupy more space than they normally do).

There are four cerebral ventricles (see *Fig. 11*). The **fourth ventricle** extends from the spinal cord to the pons. Between the fourth and third ventricles is the **aqueduct of sylvius**. The aqueduct then expands into the **third ventricle** which lies in the center of the brain, adjacent to the **thalamus** and the **hypothalamus**. From the third ventricle, the channels enter the cerebrum and bifurcate. The two channels enlarge and form the **lateral ventricles**.

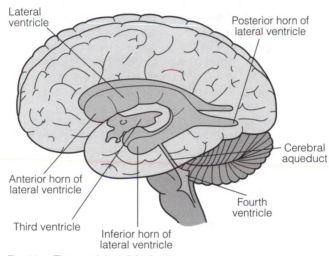

Fig. 11. The ventricles of the brain.

C2 LOCALIZATION OF CORTICAL FUNCTION

Key Notes

What is meant by localization?	Localization is the idea that functions of the brain are each concentrated within a single or a small set of brain structures.
Localization of vital functions	Vital functions such as breathing, heart rate, blood pressure, and homeostasis are all controlled by discrete structures within the brain. Within the medulla oblongata there are control centers for respiration and cardiovascular functioning, and the reticular formation is responsible for arousal and some aspects of sleep. The hypothalamus is the structure where most of our homeostatic control takes place. Sexual functioning comprises a large range of separate, but coordinated, elements, each of which is relatively localized to a small region of the brain.
Localization of higher functions	There is reasonably good evidence that functions such as language, memory and emotion are localized to a degree. However, this localization might not be as clear cut as for the vital functions. Language is a good example of the localization of sub-functions, with speech production localized in Broca's area, and comprehension in Wernicke's area. Memory is also localized to some degree. Emotion is localized in a brain system: the limbic system.
Related topic	The anatomy of the central nervous system (C1)

What is meant by localization?

It has long been maintained that certain biological and psychological functions are localized within certain structures or regions of the brain. For some functions this is very clear, as removal of the structure concerned results in the total and irretrievable loss of that function. However, for other functions there are several issues that need to be resolved. This tends to be the case for higher functions rather than vital functions.

For some higher functions, there is clearly more than one structure that has a role to play. Some of the debates about localization center on whether these are multiple centers of control or whether each region is involved in a slightly different aspect of that function. If the former were the case, it would introduce a degree of redundancy in brain functioning, which might not be a bad thing. Such redundancy could allow for the survival of that function following brain damage (see Topic C4). Alternatively, from an evolutionary perspective, a phylogenetically older part of the brain might control the function up to a level of difficulty at which more complex processing is required. A phylogenetically newer part of the brain would need to be 'up to speed' in order to take over the control of that function when required. This is suggested as one role of the hippocampus (see Topic P4) if the cerebellum cannot cope with the complexity of learning required.

Localization of vital functions

Strictly speaking, the vital functions include respiration, heart function, homeostasis and all of the other functions without which we would die. However, for some functions it is either not clear if they are vital in that sense (e.g. sleep; see Topic K2; or the control of biological rhythms; see Topic K4) or they are vital for the survival of the species rather than the individual (sexual behavior; see Topic N3). Some of these functions are localized to regions of the hindbrain. Within the medulla oblongata there are control centers for respiration and cardiovascular functioning, and the reticular formation is responsible for arousal and some aspects of sleep.

The hypothalamus is the structure where most of our homeostatic control takes place. For example, nuclei within the hypothalamus are critically involved in eating and satiety (Topic M2), drinking (Topic L2), and temperature regulation (Topic L1). In addition to its role in homeostasis, the hypothalamus has a major role in the control of biological rhythms (Topic K4). We have an internal body clock that operates on a circadian rhythm of approximately 24 hours. This biological clock synchronizes our sleep–waking cycle and also the release of a wide range of hormones. The clock is located in the suprachiasmatic nucleus (SCN) of the hypothalamus. Since destruction of the SCN completely destroys the circadian rhythm, it is safe to say that this is a clearly localized function.

Sexual functioning comprises a large range of separate, but coordinated, elements (see Topics N2 and N3). These include hormonal control, olfaction, and other sensory systems, as well as a number of higher functions. Nevertheless, we can still consider localization of function here as each of these elements is relatively localized to a small region of the brain. For example, whilst hormonal control is localized in the hypothalamus and the pituitary gland, male sexual behavior is controlled by the medial preoptic area (just rostral to the hypothalamus) and female sexual behavior is controlled by the ventromedial nucleus of the hypothalamus.

Localization of higher functions

It is harder to determine whether many of the higher functions are localized, mainly because they are multi-faceted. None are simply located in one discrete region of the brain. Nevertheless, there is good evidence that functions like language, memory and emotion are localized to a degree. The localization that does take place is usually that of sub-functions that are a part of the whole process.

Language is a good example of the localization of sub-functions (see Topic Q2). There are two main language centers. Broca's area, located in the left frontal lobe, is concerned with speech production, whereas Wernicke's area, located in the left temporal lobe, is concerned with speech comprehension.

Memory is another higher function that is localized to some degree. However, unlike language, there appear to be a number of places in the brain that are capable of storing memories (see Topic P3), albeit of varying complexities. The picture becomes less clear with a function like memory because it is not easy to clearly define what its sub-functions are. For example, the distinction between short-term and long-term memory or between episodic and semantic memory is not clear even within cognitive psychology. Nevertheless, data from neuropsychological studies have shed some light on which parts of the brain seem to have a more obvious role (see Topic P5).

Emotion can be said to have a different form of localization in the brain (see Topic O2). Here, instead of the localization being characterized by specific structures, the localization is to a brain system. This system is the limbic system and it has a role in the perception, recognition and expression of emotion. One part of the limbic system in particular, the amygdaloid complex, plays a key role in integrating all of the biological responses associated with an emotional reaction. For example, Rolls (1982) has shown that amygdala neurons respond to emotional stimuli, and Davis (1992) has shown that if the central amygdala is stimulated an animal will show fear and agitation.

Other higher-level functions are less easy to pinpoint as being localized. Perceptual processes are located partly within the region of the primary sensory area for the modality, but for the non-visual senses further processing takes place in the association cortex in a rather undefined way. Likewise, attention processes have been hard to localize in any meaningful way.

C3 HEMISPHERIC LATERALIZATION

Key Notes

What is meant by lateralization?	Lateralization is the control of a function by one hemisphere. Studies of patients with a severed corpus callosum have provided evidence that some functions or parts of functions are controlled by one hemisphere rather than the other. These include language, visual functions, detection of odor, information processing, and emotion.
The lateralization of language	The lateralization of language by the left hemisphere is not clear cut. Deficits in speech production follow damage to Broca's area in the left frontal cortex. Damage to the left temporal cortex produces language comprehension difficulties. But while 70% of left-handers have left hemisphere control of language, half of the remainder have right-sided control and half bilateral control. Prosody (intonation) is a right hemisphere function.
Split-brain studies	Studies of split-brain patients confirm that language resides in the left hemisphere of most people. The right hemisphere is dominant for a number of visual and spatial processes, for example, manipulating patterns. The left hemisphere can be regarded as playing an *analytic* role whereas the right hemisphere is seen as having a more *holistic* role.
Related topics	The anatomy of the central nervous system (C1) Localization of cortical function (C2)

What is meant by lateralization?

The cerebral hemispheres are split into two halves (left and right), but are connected by the **corpus callosum**, a massive fiber tract. The corpus callosum allows the two sides to communicate with each other. Under normal circumstances this makes it very difficult to ask the question about whether any functions are solely controlled by one side of the brain. However, brain damage caused by injury and by surgery involving the severing of the corpus callosum have provided evidence that some functions or parts of functions are controlled by one hemisphere rather than the other. The best-known lateralized function is language, but a number of others have been discovered. These include visual functions, the detection of odor, information processing, emotion and handedness. There are even sex differences in laterality although these are probably quite small according to Springer and Deutsch (1997).

The lateralization of language

It has long been known that strokes that result in language difficulties (**aphasia**; see Topic Q1) are usually accompanied by paralysis of the right side of the body. Moreover, paralysis of the left side rarely occurs together with aphasia. Since the right side of the body is controlled by the left hemisphere, this represents strong evidence that the control of language takes place there also.

Work by Paul Broca in the middle of the 19th century confirmed that deficits in speech production are accompanied by damage to a region of the left frontal cortex. Carl Wernicke, around the same time, was seeing patients with language comprehension difficulties. On post-mortem it was discovered that these patients had damage to their left temporal cortex. This seemed to be clear cut evidence that language was a left hemisphere function. However, more recent evidence has shown that language functioning is not quite so clear cut.

Sodium amobarbital can be used to anesthetize one hemisphere of the brain at a time (the **Wada test**). Using this technique, Milner (1974) found that over 95% of right-handed people have language localized in the left hemisphere. For left-handed people the picture was different but not simply the reverse. 70% of left-handers were found to have left hemisphere control of language. Of the other 30%, half had right-sided control and half had bilateral control. However, data concerning the recovery of language following a left-sided stroke suggest that maybe more left-handers have bilateral control.

Even in right-handed people it is not the case that language is totally lateralized. **Prosody** is the intonation we put on speech to give it meaning. Saying 'the greenhouse' is differentiated from saying 'the green house' by the use of prosody. Evidence suggests that prosody is a right hemisphere function (e.g. Etcoff, 1989).

Split-brain studies

Patients who suffer from epilepsy and who fail to respond to drug treatments might rarely be given a split-brain operation to prevent the epileptic focus from spreading across from one hemisphere to the other. The operation involves a complete severing of the **corpus callosum**. This is such a drastic measure that it has only ever been performed a few hundred times. Nevertheless, such patients offer a unique opportunity to study lateralization as the hemispheres can no longer communicate with each other.

Split-brain studies have confirmed the belief that language resides in the left hemisphere of most people (Gazzaniga and Sperry, 1967). If you present a word to the right visual field it will be seen by the left hemisphere (see Topic G1). The person will be able to repeat the word that they have seen. If, however, the word is presented to the left visual field so that it is seen by the right hemisphere, the person will not be able to say the word. That there is some language capability in the right hemisphere is shown by the fact that the person will, nevertheless, be able to reach for an appropriate object with their left hand. However, this seems to be restricted to single words rather than grammatical constructions.

Split-brain studies have also allowed researchers to investigate what it is that the right hemisphere actually does. It seems that the right hemisphere is dominant for a number of visual and spatial processes, for example, manipulating patterns. However, more recently it has been suggested that instead of thinking of one hemisphere being for one set of tasks (e.g. verbal) and the other being for a different set of tasks (e.g. visual), we ought to think of the hemispheres as having different roles within the same tasks. So the left hemisphere is seen as playing an *analytic* role whereas the right hemisphere is seen as having a more *holistic* role (Harris, 1978).

When it comes to a split-brain patient making decisions, which hemisphere is dominant? One might be tempted to think that it would be the hemisphere best suited to the task. However, it seems that, to some extent, each hemisphere has a mind of its own. This is not surprising as the hemispheres are unable to communicate and so they develop their own sets of perceptions, memories, and so on. Occasionally, this results in the patient trying to do opposite things at the same time (e.g. reaching for something with one hand and stopping the reaching with the other hand). For the most part, split-brain patients avoid such conflicts by using strategies to ensure that as much information as possible goes to both hemispheres (for example, by using eye movements to pass visual information to both hemispheres).

C4 BRAIN DAMAGE AND RECOVERY

Key Notes

The physiological effects of brain damage	The short-term effects of shock and edema following brain injury can suggest worse loss of function than occurs long term. Anterograde degeneration is the death of an axon separated from its cell body. If this causes death of other neurons, this is known as transneuronal degeneration. Retrograde degeneration is the death of a cell body following severance of an axon.
Mechanisms of brain recovery	In denervation supersensitivity, cells that no longer receive part of an input become hypersensitive to remaining input. In collateral sprouting, intact neurons surrounding the damaged area send out new branches from their axons. Neural plasticity describes a different region of the brain taking over the role of the damaged area. Behavioral compensation involves the individual learning new strategies to circumvent the problem caused by the brain damage.
Functional aspects of recovery	That the brain has recovered physically is not a necessary indication that it has recovered functionally.
Related topic	The anatomy of the central nervous system (C1)

The physiological effects of brain damage

Just as with any other part of the body, injury to the brain sets off a cascade of events designed to minimize the damage and start the healing process. One such event is **shock**. This is the temporary decrease in functioning of an area that follows a loss of input. Another major event is **edema**. This is the swelling that occurs around tissue following an injury. The swelling is a defense mechanism, but in the brain can cause an increase in pressure that stops previously undamaged neurons from functioning properly. This may create the early impression that the loss of function following the injury is worse than it really is. Indeed, in 1911, Von Monakow observed that functions may return quite suddenly following injury, a phenomenon he termed **diaschisis**.

As well as these initial events, the result of brain damage to one area might be the subsequent degeneration of other areas. **Anterograde degeneration** describes the fact that if an axon is separated from its cell body, the axon will die. This can even extend beyond the damaged neuron, as some neurons, if deprived of their inputs, may, themselves, die. This is referred to as **transneuronal degeneration**. Alternatively, **retrograde degeneration** results in the severed axon and its originating cell body dying.

Mechanisms of brain recovery

Unlike peripheral nerves, there is little evidence that neurons in the central nervous system can regenerate. Nevertheless, a number of mechanisms exist to allow the brain to continue to function as normally as possible. One such mechanism is **denervation supersensitivity**. This is where cells that no longer receive part of an input become hypersensitive to any remaining input that still exists. This is seen in Parkinson's disease where neurons in the basal ganglia compensate for the progressive loss of input from the substantia nigra by becoming more sensitive to the lower level of dopamine that is available. For a time, this denervation sensitivity can completely compensate for the loss but eventually progressive degeneration will result in a functional deficit.

Another recovery mechanism is **collateral sprouting**. This is where the intact neurons surrounding the damaged area send out new branches from their axons. This results in physiologically active connections to the neurons that have lost their input. Similarly, a neuron that has lost its target can send out projections until it finds a new one. This is referred to as **rerouting**.

As well as these physical mechanisms, there are a number of possible compensatory mechanisms that the brain can employ. For example, a different region of the brain that normally carries out a related function might take over the role of the damaged area. There is ample evidence to suggest that the brain is capable of exhibiting the necessary **neural plasticity**. As an alternative, some have suggested that there is a degree of **redundancy** built in just in case. The redundancy might involve more than one region of the brain normally carrying out a single function. Hence, if one region is damaged, the other can take over almost immediately.

Another compensatory mechanism is **behavioral compensation**. This involves the individual learning new strategies to circumvent the problem caused by the brain damage. So, for example, a person might become more routinized and organized so that their memory deficit is not so disabling.

Functional aspects of recovery

There is a note of caution that we must be aware of amidst all of these mechanisms of recovery. That the brain is capable of exhibiting a number of physiological solutions to the loss due to brain damage is unquestionable. However, what is questionable is the functional significance of these mechanisms. For example, imagine that a region of the brain that used to control foot movement is damaged. If that region now becomes innervated through collateral sprouting by an area that controls hand movement what will be the resulting function? Might it be that a decision to move one's hand results in the movement of both the hand and the foot? Most research suggests that recovery tends to be adaptive rather than maladaptive but there is an important role for rehabilitation in shaping functional recovery processes.

D1 THE SOMATIC NERVOUS SYSTEM

Key Notes

The components of the somatic nervous system	The somatic nervous system is the part of the peripheral nervous system that deals with the inputs and outputs from non-specialized sense organs and skeletal muscle.
Spinal nerves	The spinal nerves exit and enter the spinal cord all along its length. They serve all of the muscles of the body apart from the face.
Cranial nerves	There are 12 pairs of cranial nerves that enter and exit the brain directly from its ventral surface. These mostly feed the facial musculature but some serve more specialized functions.
Related topic	The autonomic nervous system (D2)

The components of the somatic nervous system

All of the nervous system that is not brain or spinal cord is referred to as the **peripheral nervous system**. The **somatic nervous system** is the part of the peripheral nervous system that mainly serves our skin and skeletal musculature. More precisely, perhaps, it is those parts of the nervous system that are not brain, spinal cord, nor autonomic nervous system.

The somatic nervous system carries information both *to* the central nervous system (sensory, or **afferent nerves**), and *away* from it (motor, or **efferent nerves**). In the sensory direction, the somatic nervous system is made up of those sensory nerves that innervate (i.e., provide a nerve supply to) the sense organs in the skin (e.g. touch receptors and pain receptors). In the motor direction, the somatic nervous system serves all of the skeletal muscles of our body. Skeletal muscles are those muscles over which we have direct voluntary control.

The somatic nervous system can, itself, be divided into two parts. The spinal nerves enter and leave the CNS via the spinal cord whereas the cranial nerves enter and leave the CNS via the ventral surface of the brain. These cranial nerves serve the head and the spinal nerves serve the rest of the body. The word **nerve** here means a bundle of axons (nerve fibers) that travel together from the central nervous system to another part of the body. Within the CNS a bundle of axons is referred to as a tract.

Spinal nerves

The spinal nerves enter and leave the spinal cord in a uniform manner all along its length. *Fig. 1* shows a cross-section through the spinal cord to illustrate the places where the sensory and motor neurons enter and leave the spinal cord. Sensory fibers enter the spinal cord through the dorsal root which is located in the dorsal half of the spinal cord. Motor fibers leave the spinal cord via the ventral root situated in the ventral half. One thing to note is the positions of the

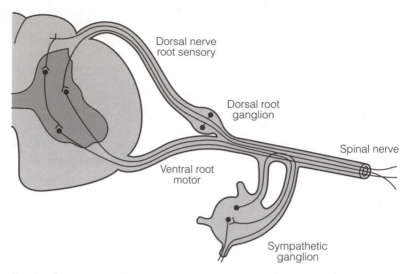

Fig. 1.   Cross-section through the spinal cord to show dorsal and ventral roots.

cell bodies. Sensory fibers have their cell bodies just outside the spinal cord, in a swelling called the dorsal root ganglion. Motor fibers, however, have their cell bodies in the ventral portion of the spinal cord itself.

Cranial nerves   There are twelve pairs of cranial nerves. *Table 1* lists the parts of the body innervated by each pair and the first thing to note is that they do not all have both sensory and motor functions. On the motor side, most but not all of the nerves serve the facial musculature. On the sensory side, there is a little more variety, with some serving the major senses. The 10th cranial nerve is called the vagus nerve and serves the internal organs rather than the head. This nerve, together with the 3rd, 7th and 9th cranial nerves, is part of the parasympathetic nervous system (see Topic D2).

Table 1. Functions of the 12 cranial nerves

	Nerve pair	Sensory	Motor
I	Olfactory	Smell	
II	Optic	Vision	
III	Occulomotor		Eye movements
IV	Trochlear		Eye movements
V	Trigeminal	Sensation of mouth and face	Mastication
VI	Abducens		Eye movements
VII	Facial		Facial movement
VIII	Auditory	Audition and balance	
IX	Glossopharyngeal	Gustation	Swallowing and vocalization
X	Vagus	Internal organs .	Internal organs
XI	Spinal Accessory		Neck movement
XII	Hypoglossal		Tongue movement

D2 THE AUTONOMIC NERVOUS SYSTEM

Key Notes

The structure of the autonomic nervous system (ANS)	The ANS serves those functions that are not under voluntary control. It has two branches. The sympathetic nervous system mainly activates systems. The parasympathetic nervous system mainly calms things down. At all preganglionic synapses of both systems the neurotransmitter is acetylcholine. At the target organ synapses the neurotransmitter substances in the sympathetic system is predominantly norepinephrine. In the parasympathetic system it is predominantly acetylcholine.
The sympathetic branch of the ANS	The sympathetic nervous system controls functions that activate the body, including the 'fight-or-flight' response. This includes causing epinephrine excretion, and diverting blood from digestive processes to muscles.
The parasympathetic branch of the ANS	The parasympathetic nervous system returns the body to its normal state once increased activities are over, and actively promotes the storage of energy. Its actions tend to be antagonistic to those of the sympathetic nervous system.
The excesses of modern living	The autonomic nervous system has evolved to cope with natural environmental events. Excessive and prolonged stress can result in ill health as a direct result of overactivity of the ANS.
Related topic	The somatic nervous system (D1)

The structure of the autonomic nervous system

The **autonomic nervous system (ANS)** is the other part of the peripheral nervous system. It has sensory and motor functions for the internal organs and glands. These tend not to be under voluntary control, and many of the muscles innervated by this system are made of smooth muscle rather than skeletal muscle. This has a different structure and different properties to the muscles that are attached to our skeleton. Like the CNS, the ANS has both afferent and efferent functions. Through these it is the main way in which the CNS controls the internal environment, and adapts the body to meet the demands placed on it by environmental changes (see homeostasis, Topic L1).

The ANS has two distinct branches, the **sympathetic nervous system** and the **parasympathetic nervous system**. In both systems the neurons that innervate the target organs leave the spinal cord and then make a single synapse before reaching the target organ. Also, in both cases the second neuron emanates from a ganglion. However, as can be seen from *Fig. 1*, the ganglion is close to the spinal cord in the sympathetic system and close to the target organ in the parasympathetic system. So the sympathetic nervous system has short **preganglionic fibers** and

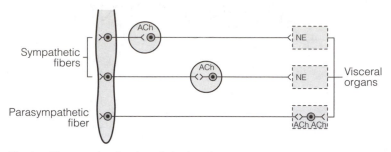

Fig. 1.   Diagram showing the relative lengths of pre- and post-ganglionic fibers in sympathetic and parasympathetic branches of the ANS.

long **postganglionic fibers** and the parasympathetic nervous system has the reverse.

At all the preganglionic synapses of both systems the neurotransmitter is **acetylcholine (Ach**; see Topic B3). However, another difference between the two systems is in the neurotransmitter substances released at the target organ synapses. In the sympathetic system, the neurotransmitter is predominantly **norepinephrine** (sometimes called noradrenaline) whereas in the parasympathetic system it is predominantly ACh. For the most part, this leads to the two functions being antagonistic where there is innervation of the same target by the two systems. In other words, if the sympathetic action is excitatory then the parasympathetic activity tends to be inhibitory.

The systems also differ with respect to where the fibers leave the spinal cord. The sympathetic fibers all emerge from thoracic and lumbar sections of the spinal cord whereas the parasympathetic fibers come from the cranial and sacral portions of the cord.

A final anatomical feature worthy of note is that transmission from the postganglionic fiber to the target organ is not always via a traditional synapse. At some target organs, the postganglionic neuron's terminal does not have a bouton and a synaptic cleft but a spread-out network of connections called a **plexus**.

The sympathetic branch of the ANS

The **sympathetic nervous system** is mainly involved in readying the body for action. A typical situation in which the sympathetic nervous system is called into action is the 'fight-or-flight' response (see Topic O1). *Table 1* shows the functions that the sympathetic nervous system has on each of its target organs. We can see how these come together as an integrated response if we consider a 'fight-or-flight' situation.

Imagine you are trekking through a forest when you chance upon a bear looking straight at you. Imagine also that you have decided not to try to fight but to run with all of the speed that your body can muster. You want your pupils to dilate so that you can see clearly any obstacles in front of you as you run away at speed. You will need to get as much oxygen as possible to your muscles so you will need to increase your heart rate and to relax your airways for deeper breathing. Increased sweating will help to keep your body cool as you run away. Your muscles will need more energy as well as more oxygen so the liver enables more glucose to be available. Finally, the sympathetic nervous system causes the hormone **epinephrine** (sometimes called adrenaline) to be released from the adrenal gland. This has a general mobilizing effect on the body.

Table 1.   Actions of the ANS at each target organ

Target organ	Sympathetic action	Parasympathetic action
Eyes	Pupil dilation and tear inhibition	Pupil constriction and tear production
Mouth	Inhibits salivation	Stimulates salivation
Lungs	Relaxes the airways	Constricts the airways
Heart	Increases heart rate	Decreases heart rate
Sweat glands	Stimulates sweating	
Blood vessels	Constricts vessels in the skin	
Liver	Stimulates glucose release	
Pancreas	Inhibits digestion	Stimulates digestion
Stomach	Inhibits digestion	Stimulates digestion
Intestines	Inhibits digestion	Stimulates digestion
Adrenal medulla	Stimulates secretion of adrenaline and noradrenalin	
Bladder	Relaxes the bladder	Contracts the bladder
Sex organs	Stimulates orgasm	Stimulates sexual arousal

If you tried to do all of this and kept your other functions going as normal then it is likely that you would run out of fuel quite rapidly. So the other part of the sympathetic response is to reduce the activity of those areas of the body not required in running away. Hence the blood vessels to the skin are constricted to allow blood to be directed to the muscles. The digestive system is inhibited as processing food is not necessary in the short term. Similarly, salivation, which is required to deal with food, is not required at this time, hence the common experience of a dry mouth at times of stress.

The parasympathetic branch of the ANS

The actions of the **parasympathetic nervous system** are also shown in *Table 1*. As you can see, the actions tend to be antagonistic to those of the sympathetic nervous system. The parasympathetic nervous system is active when we are at relative rest. If we have just successfully fought or run away then we need to bring our heart rate back down, reduce our breathing, and so on. We will also have expended a lot of energy, including extra resources that we had stored up for just such an eventuality. When we are at rest, the parasympathetic nervous system mobilizes the storage of energy via increased digestion.

The excesses of modern living

The ANS was an evolutionary solution to a particular problem, that of the danger of becoming a meal for another animal. However, in a modern context, our ANS can be activated by everyday events that lie outside of 'fight-or-flight' scenarios. For example, as you approach the exam room you will most likely experience an activation of the fear mechanism that results in an increase in sympathetic nervous activity. However, whilst your body is being readied for some physical activity, the only physical activity you are about to engage in is excessive wrist movement. If this 'fight-or-flight' response happens as a one-off, no long-term damage will be done. However, if this is repeated regularly, as it might be if you were under long-term stress, then the result could be illness or even death. It should be pointed out that the ANS is not the sole factor in this long-term stress response but it is certainly a key player (see Topic O5).

E1 HORMONES AND THEIR ACTION

Key Notes

What is a hormone?	The hormone or endocrine system provides a slower control system than the nervous system. A hormone is a chemical that is released from an endocrine gland or from other tissues. Hormones are usually secreted into the blood.
Where hormones act	Hormones act at target cells. For any given hormone, target cells are specifically receptive to that hormone. However, a target cell can respond to more than one hormone. Some hormones act on the nervous system.
Positive and negative feedback	The amount of hormone in the bloodstream is regulated by positive and negative feedback mechanisms. Negative feedback is the usual way by which the endocrine system regulates the amount of hormone circulating in the blood.
Related topic	The principles of drug action (F1)

What is a hormone?

The **hormone** or **endocrine system** provides, generally, a slower means of control over the functions of the body than the nervous system. Like neurotransmitters (see Topic B3), hormones are specialized chemicals that change the activity of cells by attaching to receptors on the cells. To emphasize the close relationships we will see between hormones and the actions of the nervous system, the endocrine system is sometimes referred to as the **neuroendocrine system**. Most hormones are released from specialized endocrine glands, and the rest from special cells within organs like the kidney and stomach. Wherever the hormone is released from, it usually travels to its target through the bloodstream.

Hormones fall mainly into four different groups of chemicals. These are:

- peptides and proteins;
- amino acids;
- fatty acids;
- steroids.

The peptides and proteins exhibit a wide variety of sizes and shapes of molecule and include insulin (see Topic E7). The amino acids include the thyroid hormones such as thyroxine (Topic E5), and epinephrine (adrenaline; Topic E3), which is derived from tyrosine. The fatty acids include the prostaglandins (Topic I1) and are made from polyunsaturated fats. Finally, the steroids include testosterone (Topics E6 and N2) and are derived from cholesterol. These subdivisions are sometimes simplified into steroids and non-steroids.

Where hormones act

Hormones mostly travel in the bloodstream where they come into contact with all cells. However, they only have an effect once they reach their own **target cells**. These cells have specially adapted receptors that accept the hormone in the same 'lock and key' fashion that we saw for neurotransmitters (Topic B3). A target cell may be receptive to just one particular hormone or to a number of different hormones. For each hormone, though, there is a specific receptor site.

endocrine action:
    distant

paracrine action:
    localized

autocrine action:
    cells that
    released it

Most hormones travel in the bloodstream to affect a distant target organ. However, some act more locally, or even on the very cell that secreted it. To differentiate between these actions the term **endocrine action** is used to describe a distant action, **paracrine action** is used to describe a more localized action, and **autocrine action** is used to describe a hormone that acts on the cells that released it.

Another noteworthy feature of hormones is their speed of action. When a neuron fires, its effects are virtually immediate as conduction is fast and the neurotransmitter substance is, to all intents and purposes, instantaneously released. By contrast, when a hormone is released it must usually travel in the bloodstream until it reaches its target destination. Add to this the fact that the release of the required hormone is sometimes only triggered by another hormone also released into the bloodstream, and the time it takes for the required hormone to finally exert its effect on the target organ can be of the order of a few minutes. Compare, for example, the speed with which you remove your hand from a hot object (neuronal) with the time it takes to induce milk ejection from a stimulated nipple (30–60 seconds).

Direct integration between endocrine activity and activity of the nervous system is achieved by a group of chemicals called neurohormones. These are released by the endocrine system but have their targets in the nervous system. For example, **cholecystokinin** (CCK) is released by the small intestine and acts on certain brain stem nuclei. Similarly, epinephrine is released into the bloodstream by the adrenal gland and travels to the neurons of the sympathetic nervous system where it causes the sympathetic arousal described in Topic D2.

Positive and negative feedback

The amount of a hormone that is present in the blood is crucial for the effective working of the endocrine system. Hormonal action needs a monitoring system so that the amount of hormone can be increased or reduced as necessary. Most hormonal control is by **negative feedback**. *Fig. 1* illustrates how this form of

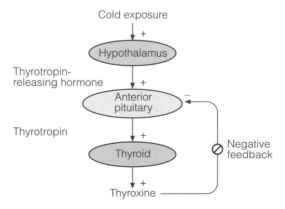

Fig. 1.   Diagram of a negative feedback system.

feedback is usually achieved (see also Topic L1). The cells that initially release the hormone (or sometimes an activating chemical; see Topic E2) are receptive to the presence of the hormone in the blood. As more and more of the hormone is released, the quantity in the blood rises. At some cut-off point, the receptors within the endocrine gland register that there is enough hormone circulating and turn off production and/or release of any more hormone. For some hormones, there is more than one place at which such monitoring occurs.

In a few cases, the feedback system is positive. Here, the detection of the presence of the hormone in the blood triggers even more to be released. An example of this is the hormone **oxytocin**, released from the posterior pituitary gland (see Topic E2). This stimulates milk secretion from the breast. The act of suckling stimulates the release of oxytocin and is a positive feedback mechanism.

E2 THE PITUITARY GLAND

Key Notes

Structure of the pituitary gland	The pituitary gland has two main parts, anterior and posterior. Both have a vascular link with the hypothalamus, and the posterior pituitary has a neural connection with it. They are responsible for the release of different hormones and stimulating hormones.
The pituitary gland as a master gland	Most of the hormones that circulate around the body are controlled in one way or another by the pituitary gland.
The role of the hypothalamus	The hypothalamus controls hormonal release. It does so mainly by influencing the release of hormones and stimulating hormones from the pituitary gland. However, control of the anterior pituitary is by hormones released by the hypothalamus but the posterior pituitary is controlled by neuronal input from the hypothalamus.
Hormones released by the posterior pituitary gland	The posterior pituitary gland secretes just two hormones, oxytocin and vasopressin. The former is involved in lactation and the latter is involved in blood pressure control.
Hormones released by the anterior pituitary gland	Most of the hormones released by the anterior pituitary are stimulating hormones. These stimulate other glands to release their hormones into the blood. The exception here is prolactin which is a hormone that acts directly on target cells.

Structure of the pituitary gland

The pituitary gland is actually two main separate glands, an anterior gland and a posterior gland (*Fig. 1*). It is located at the base of the skull. The two parts control the release of different hormones and are controlled differently by the hypothalamus. The **anterior pituitary gland** is also known as the **adenohypophysis** and the **posterior pituitary gland** is also known as the **neurohypophysis**. These names reflect the different way in which the hypothalamus controls secretion from these glands.

The pituitary gland hangs down under the main part of the brain and is connected to the **hypothalamus** by the **pituitary stalk**. The stalk contains a rich mix of blood vessels and neurons. The neurons only extend into the posterior pituitary. Blood vessels supply both parts of the pituitary gland but those vessels going to the anterior pituitary contain chemicals that control its functioning. The blood vessels to the posterior pituitary only perform the normal functions supplied by blood vessels anywhere (nutrition, oxygen, etc.).

The pituitary gland as a master gland

The hormones released by the pituitary gland have such far-reaching consequences that the gland has been referred to as the 'master gland'. Many other glands in the body secrete their hormones in response to hormones released by

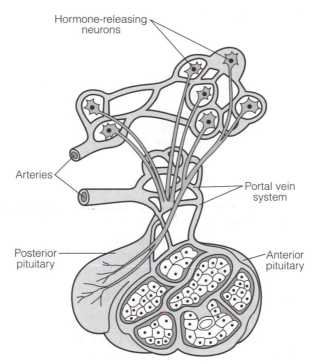

Fig. 1.   Structure of the pituitary gland.

the anterior pituitary gland. This regulation occurs because the pituitary gland secretes a number of chemicals that are not really hormones in their own right but are referred to as **stimulating hormones**. These stimulating hormones then stimulate the release of hormones from other glands. This system serves some of the negative feedback loops described in Topic E1.

The role of the hypothalamus

The hypothalamus plays an important role in controlling the release of hormones. It has direct control over the release of many pituitary hormones and also contains a number of detectors for circulating hormones. As already mentioned, the hypothalamus sends neuronal projections to the posterior pituitary gland. These come from two hypothalamic nuclei called the **supraoptic nucleus** and the **paraventricular nucleus**. The hormones released by the posterior pituitary are actually synthesized in the hypothalamic neurons and are transported down the axon to the posterior pituitary. Once there, the release of these hormones is controlled by neuronal activity in these hypothalamic nuclei.

The connection between the hypothalamus and the anterior pituitary is by way of a rich vascular system called the **hypothalamic–pituitary portal system**. Neuronal cells in the hypothalamus are called **neurosecretory cells** because they produce and release hormones that are secreted directly into the blood from the ends of their axons. Many of the hormones released by the hypothalamus are called releasing hormones as they, in turn, cause the release of hormones from the anterior pituitary. Examples of these are **corticotropin-releasing hormone (CRH)** and **growth-hormone-releasing hormone (GRH)**. A full list together with their actions at the anterior pituitary is given in *Table 1*.

Table 1.   Hypothalamic hormones and their effect on the anterior pituitary gland

Hypothalamic hormone	Effect on anterior pituitary
Corticotropin-releasing hormone (CRH)	Secretion of ACTH
Thyrotropin-releasing hormone (TRH)	Secretion of thyrotropin
Growth-hormone-releasing hormone (GRH)	Secretion of growth hormone
Somatostatin	Inhibits secretion of growth hormone
Gonadotropin-releasing hormone (GnRH)	Secretion of LH and FSH
Prolactin-releasing hormone (PRH)	Secretion of prolactin
Prolactin-release inhibiting hormone (PIH)	Inhibits secretion of prolactin

Hormones released by the posterior pituitary gland

The posterior pituitary gland releases two hormones into the blood. These are oxytocin and vasopressin. **Oxytocin** has two main functions:

- stimulating contractions of the uterine muscles during childbirth;
- the **milk letdown reflex** (see Topic N4). When an infant starts to suckle, cells in the skin inform the hypothalamus. This then stimulates the release of oxytocin from the posterior pituitary, increasing milk production. This is a positive feedback loop (see Topic E1). Once the infant is sated, the stimulation to the hypothalamus will cease and the release of oxytocin will stop.

Vasopressin is also known as **antidiuretic hormone** (**ADH**). As we will see in Topic L2, ADH is produced by cells in the hypothalamic nuclei in response to thirst, and increases blood pressure through constriction of small blood vessels.

Hormones released by the anterior pituitary gland

Six main hormones are released by the anterior pituitary. Two of these influence the gonads, two cause release of hormones from other glands, and two have direct actions themselves. The two gonad-influencing hormones are **luteinizing hormone** (**LH**) and **follicle-stimulating hormone** (**FSH**). In females, LH stimulates the release of eggs from the ovaries and prepares the wall of the uterus for receiving the egg if it is fertilized. In males, this hormone stimulates the cells of the testes to produce testosterone. FSH stimulates the release of estrogen in females and testosterone in males (see Topic N2). It also has a role in the production of both eggs and sperm (see Topic E6).

Adrenocorticotropic hormone (**ACTH**) controls both the production and release of hormones from the adrenal cortex. **Thyroid-stimulating hormone**, as its name suggests, causes the release of thyroid hormone from the thyroid gland.

Prolactin has a role in promoting lactation (milk production) in females, and a role in maternal behavior (see Topic N4). **Growth hormone** (**GH**), or **somatotropin**, works throughout the body to promote the growth of cells. Mostly, this is via an effect on protein metabolism. The hypothalamic influence of growth hormone release via growth-hormone-releasing hormone (also known as somatocrinin) was described earlier. However, the release of growth hormone can also be prevented by the release of **somatostatin** from the hypothalamus.

E3 THE ADRENAL GLANDS

Key Notes

Structure and location of the adrenal glands	The adrenal glands are located above the kidneys. They are composed of the adrenal cortex and the adrenal medulla.
Hormones released by the adrenal medulla	The adrenal medulla secretes epinephrine and norepinephrine into the blood. These are released as part of the stress response.
Hormones released by the adrenal cortex	The adrenal cortex secretes three different kinds of hormones, the glucocorticoids, the mineralocorticoids and the sex steroids. The corticoids are involved in metabolism and the sex steroids are involved in secondary sexual characteristics.

Structure and location of the adrenal glands

The adrenal glands are located above the kidneys and consist of an outer layer called the **adrenal cortex** and an inner core called the **adrenal medulla**. The medulla is made up of chromaffin cells that release **epinephrine** and **norepinephrine** when stimulated. The stimulation comes from the lesser splanchnic nerve of the sympathetic nervous system. The cortex lies around the outside of the medulla and contains various glandular cells that secrete hormones involved with either metabolic regulation or secondary sexual characteristics.

Hormones released by the adrenal medulla

The adrenal medulla secretes epinephrine and some norepinephrine in response to sympathetic nervous system activity. Their secretion has an almost immediate effect on heart rate, which rises sharply. Other, less immediate, effects are an increase in blood pressure, sweating, a dry mouth, and an increase in energy to areas of need like the muscles. All of these are responses to 'fight or flight' scenarios and help ready the person for the likely activity (see Topics O1 and O4). More importantly, they help to sustain this readiness after the initial neuronal ANS effect.

Hormones released by the adrenal cortex

The adrenal cortex secretes a large number of hormones. These can be categorized into three groups, the **glucocorticoids**, the **mineralocorticoids** and the **sex steroids**. The main glucocorticoid is **cortisol** (**corticosterone**). Cortisol works on all the cells in the body. It increases the level of glucose in the blood, inhibits the uptake of glucose by all cells other than those in the brain, and causes the breakdown of proteins into their constituent amino acids. The increased blood glucose can be used by muscles if necessary, the brain needs an increased supply of energy, and the freed amino acids can be used for tissue repair. Cortisol is also part of the stress response (see Topic O5).

The main mineralocorticoid released by the adrenal cortex is **aldosterone**. Aldosterone controls the levels of minerals in the body by monitoring their levels in the blood and regulating their reabsorption by the kidneys. This is achieved mainly through the monitoring of blood potassium levels. When blood potassium levels rise too high, aldosterone is released and causes the reabsorption of salts by the kidneys. Through this mechanism, aldosterone plays a role in thirst and drinking (see Topic L2).

sex steroid: androstenedione

The main sex steroid released by the adrenal cortex is **androstenedione**. This contributes to the male and female patterns of body hair. In particular, it leads to the growth of pubic and axillary (underarm) hair at puberty in females and to the growth of pubic, axillary and facial hair at puberty in men. Androgens produced by the adrenal cortex probably play a role in sexual arousal in women (see Topic N2).

E4 THE PANCREAS

Key Notes

Structure and location of the pancreas	The pancreas is located at the back of the abdomen. It has cells called Islets of Langerhans that secrete its hormones into the blood.
Hormones released by the pancreas	The pancreas secretes insulin and glucagon to regulate the levels of blood glucose and the storage of fats. Somatostatin is also secreted, and helps to control the secretion of insulin and glucagon.
Related topic	Digestion, energy use and storage (M1)

Structure and location of the pancreas

The pancreas is a large gland located just below the stomach. It contains two types of cell, the **Islets of Langerhans** and the **acini**. As well as its hormonal action, it releases a large number of digestive secretions into the small intestine. These are secreted from the acini. The Islets of Langerhans are themselves made up of two types of cell, alpha and beta cells. The alpha cells secrete glucagon and the beta cells secrete insulin.

Hormones released by the pancreas

insulin

A major hormone secreted by the pancreas is **insulin**. Insulin is involved in glucose metabolism and helps glucose to enter cells (see Topic M1). Glucose cannot diffuse through cells by itself and is helped across the membrane by carrier molecules. Insulin increases the ability of the carrier molecules to move glucose into the cell.

A second action of insulin is to promote fat storage. When glucose enters fat cells it causes a set of reactions that culminate in the production of fat. Hence, by increasing the transport of glucose into cells, insulin indirectly promotes the production of fat. Conversely, in times of high energy need, glucose is diverted to those parts of the body that require it most. Without glucose entering the fat cells, the stored fat is broken down and converted into fatty acids, which can then be used as an alternative source of energy (Topic M1).

A final action of insulin is to promote the storage of glycogen by the liver. As with the storage of fat, this is an indirect consequence of insulin moving glucose into cells. When the glucose enters liver cells, about two-thirds of it gets converted into glycogen. This can then either be stored in the liver or transported to fat cells for later use.

glucagon

The other major hormone secreted by the pancreas is **glucagon**. This causes glycogenolysis, the breakdown of glycogen into glucose in the liver, allowing glucose to enter the bloodstream.

The release of insulin and glucagon are controlled by the level of glucose in the blood. The detection is done by the Islets of Langerhans cells, and when the level of blood glucose becomes too high, insulin is secreted. Conversely, when the level of blood glucose drops too low, glucagon is secreted. A secondary mechanism for controlling insulin release comes from the parasympathetic vagus nerve. This action is caused by food in the mouth. It enables insulin to be released in readiness for the glucose that will be absorbed by digestion.

The best-known disease associated with the functioning of the pancreas is **diabetes**. This occurs when the pancreas fails to secrete insulin. This causes the level of blood glucose to rise as the glucose cannot easily be transported into cells. The excess glucose is then excreted by the kidneys. As a consequence of this lack of glucose metabolism, stored fats are used as a source of energy and the person starts to lose weight. In severe circumstances a diabetic coma can ensue due to the build-up of acidic breakdown products from using large amounts of fat for energy. Unless treated quickly, a person will die in a few hours. The treatment is to administer large amounts of glucose. Patients with severe diabetes have to inject themselves with insulin daily to prevent such a coma from occurring.

Somatostatin is secreted from the pancreas. It seems to have a paracrine function here, in that it suppresses secretion of both insulin and glucagon.

E5 THE THYROID AND PARATHYROID GLANDS

Key Notes

Structure and location of the thyroid and parathyroid glands	The thyroid gland is located in the throat just below the Adam's apple. It has a butterfly shape and secretes two hormones that regulate metabolism and have an effect in nearly all parts of the body. The parathyroid gland is located in the front of the neck, close to the thyroid gland. It is actually made up of four small glands.
Hormones released by the thyroid and parathyroid glands	The thyroid gland secretes thyroxine and triiodothyronine (involved in metabolism), and calcitonin (involved in calcium regulation). The parathyroids secrete parathyroid hormone, which is responsible for regulating the levels of calcium and phosphate in the blood.
Related topic	Weight control and its disorders (M4)

Structure and location of the thyroid and parathyroid glands

The thyroid gland is a large butterfly-shaped structure that lies in the throat just below the Adam's apple. In some animals there are two glands but in humans these have become fused into one. The two lobes are connected by a narrow section called the isthmus. The thyroid gland contains epithelial cells that synthesize the main thyroid hormones, thyroxine and triiodothyronine. There are also C cells that secrete calcitonin.

The parathyroid gland is four separate nodules that protrude from the thyroid gland. The nodules consist of densely packed cells that are clustered around an abundant supply of blood capillaries.

Hormones released by the thyroid and parathyroid glands

Thyroid hormones (**thyroxine** and **triiodothyronine**) probably have an influence on all the cells of the body. They can affect development, growth and metabolism. They increase the metabolic rate and, in doing so, increase body heat. They are a cause of the metabolism of lipids (fats), producing an increase in fatty acids in the blood, and also lead to carbohydrate metabolism, which releases more glucose into the blood.

The growth action of the thyroid hormones is evident from the fact that children who have a thyroid deficiency have stunted growth. Likewise, thyroid hormones are essential for the normal development of the neonatal brain. In addition, they increase heart rate, allow us to maintain our concentration levels, and are important in the normal development of our reproductive system.

Hypothyroidism occurs when too little thyroid hormone is being secreted. If not quickly treated in a newborn baby, the likely outcome will be cretinism. If hypothyroidism develops later in life, perhaps as a result of too little iodine in the diet, the thyroid gland swells to form a characteristic goiter. **Hyperthyroidism** is when there is too much thyroid hormone being secreted. This can lead to Graves Disease, which is an immune disorder.

Calcitonin is secreted from C cells in the thyroid gland and is involved in calcium and phosphate metabolism. The hormone released from the parathyroid gland is **parathyroid hormone** (also called **parathormone**). Like calcitonin, parathyroid hormone controls the levels of calcium and phosphate in the blood, and controls the absorption of calcium during digestion.

E6 THE GONADS

Key Notes

Structure and location of the ovaries and testes	The ovaries are oblong organs that lie in the pelvis at the ends of the fallopian tubes. As well as egg production, they secrete a number of hormones. The testes are glandular organs that are suspended in the scrotum. As well as producing sperm, they secrete a number of hormones.
Hormones released by the ovaries	The ovaries produce estrogens and the progestins. Estrogens are involved in development, especially of the reproductive system. They are not exclusively female hormones. Progestins are concerned with reproduction itself.
Hormones released by the testes	The testes secrete androgens, and in particular, testosterone. Testosterone is an important hormone for sexual development in the male.
Related topics	Reproduction and sexual differentiation (N1) Hormonal control of sexual behavior (N2)

Structure and location of the ovaries and testes

The **ovaries** lie in the pelvis each side of the uterus at the ends of the fallopian tubes. Each ovary is about the size and shape of an almond. It is attached to the pelvic wall and to the uterus by ligaments. The ovary has an outer cortex and an inner medulla. In the cortex are the granulosa and theca cells that are responsible for producing the ovarian hormones.

The **testes** are egg-shaped and about 4 cm long. In order to maintain spermatazoa at a temperature slightly lower than body temperature, the testes are outside the body cavity in the scrotum. Between the seminiferous tubules that produce sperm lie the interstitial cells (also called Leydig cells) that produce testosterone.

The relationship between the estrogen produced by the ovaries and the testosterone produced by the testes is a very close one molecularly. They differ by only one molecule, and testosterone is the precursor to estrogen. Within the ovaries, testosterone is converted to estrogen by **aromatase**.

Hormones released by the ovaries

The ovaries produce two main classes of hormone: **estrogens** and **progestins**. In humans, the estrogen is **estradiol**, and the progestin is **progesterone**. Estrogen is responsible for the development of the female genitalia, the growth of breasts, the alterations of fat deposits that define the female shape, and the growth of the uterine lining (see Topic N1).

Estradiol also plays a role in the **menstrual cycle** (see also Topic N2). There is an intricate interplay between the estradiol released by an ovary producing an egg and the release of pituitary hormones. Early in the menstrual cycle, the anterior pituitary releases follicle-stimulating hormone to start the development of an egg from a **follicle** in the ovary. As the egg develops, it releases estradiol,

[handwritten margin note: estrogens: estradiol; progestins: progesterone]

which causes growth of the uterine lining but also causes a rise in the level of estradiol in the blood. Once the blood level of estradiol rises above a certain point the anterior pituitary is stimulated to release **luteinizing hormone**. The surge of luteinizing hormone leads to ovulation.

The ruptured ovarian follicle left behind after ovulation becomes a **corpus luteum**. This now produces progesterone as well as estradiol. The progesterone promotes pregnancy by maintaining the uterine lining and preventing another ovarian follicle from developing into an egg. If the egg does not become fertilized, the corpus luteum will stop producing these hormones and menstruation will ensue. If fertilization does occur then progesterone continues to be secreted and continues to maintain the lining of the uterus.

A counterintuitive feature of estrogen is that it has a *masculinizing* effect on the fetal brain (see Topics N2 and N4). Since estrogen is released into the blood, why does the female brain not become masculinized? The answer is a protein called α-**fetoprotein** that binds to the estrogen and prevents it entering the brain.

Hormones released by the testes

The hormones secreted by the testes are called **androgens**. The main hormone secreted is **testosterone**, but **dihydrotestosterone** is also secreted. Testosterone has a range of functions from the development of primary sexual characteristics in the fetus (Topic N1), to sex drive in the adult male (Topic N2). Testosterone also leads to the development of a masculine brain (Topic N3). Testosterone enters brain cells where it is converted to estrogen by aromatase. This then has a masculinizing effect on the brain.

At puberty, testosterone leads to the production of sperm, and the development of the male **secondary sex characteristics**. It is a major determinant of sex drive in males, and probably also in females (Topic N2).

E7 THE OTHER GLANDS

Key Notes

The thymus	The thymus hormone is responsible for cellular immunity. Most of its job is done shortly before birth and for a short period afterwards.
The kidneys	The kidneys secrete three different hormones; renin, erythropoietin and 1,25-dihydroxy-vitamin D3. These have a role in blood pressure, erythrocyte production and calcium balance respectively.
The pineal gland	The pineal gland secretes melatonin. This has been strongly linked with the sleep–waking cycle. It also has a role in seasonal breeding and sexual maturity.
The gastrointestinal tract	There are several hormones secreted by the gastrointestinal tract. They include gastrin, secretin, cholecystokinin, gastric inhibitory peptide and somatostatin.
Related topics	Thirst and drinking (L2) Physiological mechanisms in eating Biological rhythms (K4) (M2)

The thymus

The **thymus hormone** (also called **thymosin**) is responsible for the development of thymic lymphocytes (T-cells). These are part of the immune system and give rise to cellular immunity. Most of the production of these lymphocytes occurs before birth and in the few months or years after birth. If the thymus gland were to be removed after this period, cellular immunity would not be greatly impaired.

The kidneys

The kidneys secrete three separate hormones. Renin is secreted in response to a decrease in blood pressure. The lowered blood pressure is detected by the decreased blood flow through the kidneys. Renin acts as an enzyme to turn a plasma protein into **angiotensin**. This causes vasoconstriction, thereby raising the blood pressure back towards its normal value (see Topic L2). The other kidney hormones are concerned with red blood cell production and calcium metabolism.

The pineal gland

The pineal gland secretes **melatonin** almost exclusively at night. In humans it has a role in matching body rhythms to the day–night cycle (see Topic K4).

Another role of melatonin is in the regulation of seasonal breeding and sexual maturity. In animals that breed seasonally, the fact that melatonin is secreted at night becomes a marker for the time of year. The excessive amount of melatonin secreted during the long nights of winter lowers the level of secretion of gonadotropic hormones and thereby decreases the drive for sex. It may also have a role in halting the triggering of sexual maturity, as destruction of the pineal gland before puberty can hasten its onset.

The gastrointestinal tract

The gastrointestinal (GI) tract secretes a number of different hormones, most of which are involved in controlling digestion. Many of them also play a role in satiety mechanisms in eating (see Topic M2). **Gastrin** is secreted in response to certain foodstuffs, like proteins, being present in the stomach. It leads to an increase in the amount of acid being secreted from the stomach wall. If the pH of the stomach becomes very acidic (less than pH 3), then gastrin secretion is turned off.

Secretin is released in response to there being too much acid in the small intestine. In the presence of excess acid, secretin causes the pancreas to secrete a bicarbonate-rich fluid into the first part of the intestine to neutralize the acidity.

Like secretin, **gastric inhibitory peptide** is secreted in response to an excess of acid in the small intestine. However, it also has a secondary role in that it can cause the release of insulin from the pancreas in response to the presence of glucose.

Cholecystokinin (CCK) is secreted from the first part of the small intestine. It responds to food being present in the small intestine and causes the pancreas to release its digestive enzymes. It also stimulates the release of bile from the gall-bladder. CCK release ceases when the food molecules that led to its secretion have been digested.

Somatostatin is secreted from a large number of places, including the small intestine. In the gastrointestinal tract, somatostatin inhibits the secretion of gastrin, cholecystokinin and secretin. It also suppresses secretion of gastric acid and pepsin and so has a general effect of decreasing the rate of nutrient absorption.

F1 THE PRINCIPLES OF DRUG ACTION

Key Notes

Drug administration	The amount of a drug that is needed to have the 'desired' effect varies according to a number of factors. These include the mode of administration and the metabolic and excretion dynamics of the drug.
Factors affecting drug efficacy	How effective a drug is can depend on the dose, the potency, the mode of administration, and even the initial psychological state of the person.
The placebo effect	The placebo effect is where a person behaves as though they have been administered a drug when, in fact, they have been given something that is completely neutral.

Drug administration

The effect that a drug has depends upon a number of different factors. Among these are the mode of administration, the way in which the drug is metabolized and the speed at which it is broken down and excreted.

Drugs are usually administered through one of three methods; *oral*, *intravenous* or *intramuscular*. Other routes include subcutaneous and intraperitoneal but these are less frequently used. Intravenous administration tends to have the fastest action and oral administration the slowest. This is because drugs taken orally have to be absorbed from the gut before they can act. However, the speed of action will also depend on whether the drug is water soluble or lipid (fat) soluble. In general, psychoactive drugs tend to be fat soluble and to act faster, as they enter cells through the lipid layers of cell membranes.

Some drugs are metabolized by the body very quickly whereas other drugs are metabolized more slowly. The rate of metabolism can be affected by a number of things, such as how much has recently been eaten. Furthermore, the way in which the drug is metabolized will lead to different outcomes. If the active drug is the result of ingesting a precursor chemical that is then metabolized, then a rapid rate of metabolism can speed up the effectiveness of the drug. However, if metabolic processes break down the active ingredient then slow metabolism may be desirable.

Some drugs are not broken down within the body but are excreted from the body by the kidneys (e.g. the hallucinogenic mushroom *Amanita muscaria*). The **clearance rate** defines how quickly the drug is excreted. Again, the rate of clearance will affect how long-lasting the drug action is. Putting all of these features together means that the dynamics of drug action are complex.

Factors affecting drug efficacy

Some drugs are better at having their desired effect than others. The relationship between the dose of drug and the effect observed is called the dose-response function. For some drugs, there is an optimal dose range that has therapeutic effects. This range is called the **therapeutic window**. For other drugs the effect changes as the dose changes. So, for example, amitriptyline is a good analgesic at low doses but is an antidepressant at high doses.

In order to characterize dose effects, pharmacologists talk of the threshold dose, the ED50 and the LD50. The **threshold dose** is the smallest dose that has an observable effect. The **ED50** is **median effective dose**, which means the dose at which 50% of people will experience a therapeutic effect. The **LD50** is the **median lethal dose**, which means the dose at which 50% of people will die. The LD50 is calculated using animal trials and so it is always an estimate.

We can also consider efficacy in terms of potency. **Potency** refers to the relative ability of a drug to achieve a given effect. A smaller dose of morphine is needed to achieve a certain level of analgesia than the dose of aspirin needed to achieve the same analgesic effect. Hence, even though both drugs belong to the same class, morphine is a more potent drug than aspirin.

The placebo effect

If you are told that you are being given a drug for a particular reason then you will have an expectation that the drug will have a particular effect. It is likely that, under such circumstances, you will experience an effect even if the actual substance you have been given is a sugar pill. This is known as the **placebo effect**. The existence of the placebo effect means that in order to properly characterize the effects of a drug, some participants must, unknowingly, be given the drug and some must, unknowingly, be given a placebo. (Ideally, the people carrying out the test do not know who is receiving the drug. This is a **double blind trial**.) Comparisons can then be made as to the relative effects of the drug and the placebo. Only those effects shown by the people given the drug and not shown by the people given the placebo can truly be called effects of the drug.

F2 SITES OF DRUG ACTION

Key Notes

Presynaptic effects	Drugs can act on the synthesis or storage of a neurotransmitter. They can also affect its release into the synaptic cleft.
Effects on receptors	Drugs can affect the way in which a neurotransmitter interacts with the receptor. This can be as an agonist, mimicking the effect of the neurotransmitter, or as an antagonist, blocking it either temporarily or permanently.
Effects on reuptake and degradation	Some neurotransmitters are taken back up into the presynaptic cell (reuptake) and some are broken down in the synaptic cleft (degradation). They may also be broken down after reuptake. Drugs can affect any of these processes.
Related topic	Synaptic transmission (B3)

Presynaptic effects

All psychoactive drugs work on some part of the synapse (see Topic B3). It is important to note at the outset that drugs are not classified according to where they act (see Topic F3). However, a consideration of drug action from the perspective of the synapse can help the appreciation of how drugs in the same class can work at different parts of the synapse but still achieve the same end result. It can also help in the appreciation of differences in side effects. Drugs producing **presynaptic effects** affect the synthesis, storage or release of a neurotransmitter substance.

Drugs like **p-chlorophenylalanine** affect the *synthesis* of 5HT (serotonin) by blocking the enzymatic reaction that converts tryptophan into 5-hydroxy-tryptophan (a precursor to 5HT). Incidentally, in the 1970s a number of papers came out suggesting that p-chlorophenylalanine might be an aphrodisiac. More recently, it has been investigated for its possible link with the effects of **methylenedioxymethamphetamine** (**MDMA**), also known as ecstasy (see Kish, 2002).

Another presynaptic effect can be on the *storage* of the neurotransmitter. The classic example here is **reserpine**. This drug depletes catecholamine neurons of their neurotransmitter. The reserpine is taken up by the synaptic vesicles, and causes them to release their neurotransmitter, which then leaks into the synaptic cleft. This leads to a biphasic response upon reserpine administration. The first effect is an enhancement of the natural neurotransmitter effects because of this leakage of neurotransmitter. However, once all of the neurotransmitter has been used up, the subsequent effect is one of depletion; that is there is a drop in synaptic transmission.

The final stage of presynaptic action is *release* from the terminal bouton. For example, dopamine release can be blocked by **gamma-hydroxybutyrate** (**GHB**). GHB acts particularly on the dopamine cells of the basal ganglia. It is becoming popular at 'raves' but has been labeled as a date-rape drug. On the other hand,

amphetamine causes an increase in the release of epinephrine and, to a lesser extent, dopamine.

Effects on receptors

At the receptor site a drug can have one of three effects. It can _mimic_ the naturally occurring neurotransmitter and thereby produce the same effects. Such a drug is called an **agonist**. An example is **lysergic acid diethylamide (LSD)** which has a shape that is similar to serotonin. When LSD comes into contact with the receptor, the same cascade of events is triggered as if serotonin had attached to the receptor.

A second action that a drug can have is to _block_ the receptor in a non-permanent way. Such a drug is called a **reversible antagonist** and an example is **haloperidol**. Haloperidol blocks dopamine receptors and is therefore used as an antipsychotic drug (see Topic R1). However, dopamine and haloperidol will fight for the occupation of the receptor sites. If the amount of dopamine increases and the amount of haloperidol decreases then dopamine will again occupy the receptor sites. The relative strengths of a drug and the naturally occurring neurotransmitter will depend, in part, on the relative affinity the receptor has for each chemical.

Some chemicals are highly toxic poisons because they combine with the receptor site in a permanent way. These drugs are **irreversible antagonists**, an example of which is the snake venom **alpha-bungarotoxin** (from the Many Banded Krait). This binds to nicotinic receptors (the acetylcholine receptors at the neuromuscular synapse) but does not cause any postsynaptic effect. Since the chemical cannot then be removed from the receptor, the naturally occurring acetylcholine (ACh) cannot act. The result is that the voluntary muscles become paralyzed. When this happens to the muscles required for breathing, death ensues.

Effects on reuptake and degradation

Once the neurotransmitter has detached from the receptor site, its effect is terminated by either _degradation_ within the synaptic cleft or _reuptake_ by the presynaptic cell. **Acetylcholinesterase (AChE)** is the enzyme that breaks down (degrades) ACh. Drugs like **physostigmine** reversibly bind to AChE and stop it working. The result is that the action of ACh is prolonged. Toxic chemicals like DFP (a nerve agent) can bind irreversibly to AChE. This obviously leads to a permanent increase in the effect of ACh release.

Drugs can also block reuptake of the neurotransmitter, and the group of antidepressants called the tricyclics work in this way (see Topic R2). These drugs include **desipramine,** which blocks the reuptake of norepinephrine, and **imipramine** and **amitryptyline,** which block the reuptake of serotonin. New drugs like **fluoxetine** (Prozac™) are selective serotonin reuptake inhibitors (SSRI drugs).

Finally, drugs can affect the breakdown of neurotransmitter once it has been taken back up into the presynaptic cell. Of particular interest here are antidepressant drugs like the **monoamine oxidase inhibitors (MAOIs**; see Topic R2). The enzyme monoamine oxidase breaks down norepinephrine and dopamine in the synaptic cleft and in the presynaptic terminal. However, it only breaks down serotonin within the presynaptic terminal. The MAOI **iproniazid** prevents this breakdown of serotonin within the presynaptic terminal. The result is that the free-floating serotonin leaks back out into the synaptic cleft where it attaches again to the receptors, continuing the activity of the synapse.

F3 PSYCHOACTIVE DRUGS

Key Notes

Classes of psychoactive drug	Drugs can be classified according to the type of action that they have on our psychological functioning. There are five main classes of drugs.
Stimulants	Stimulants cause a general increase in brain and behavioral activity. They include caffeine, nicotine, amphetamines and cocaine. Stimulant drugs have various modes of action, and differing effects.
Depressants	Depressants cause a general decrease in brain activity. They include the minor tranquilizers such as the barbiturates and benzodiazepines, and alcohol. All depressants act by reducing activity at GABA synapses. Alcohol initially has a stimulating effect as it acts first to depress inhibitory mechanisms. Alcohol and barbiturates have side effects and a high potential for dependence.
Psychedelics	These are drugs that cause hallucinations. They include LSD, cannabis, certain varieties of mushroom, and perhaps ecstasy. Their effects differ, and they have undesirable side effects.
Opiates	The opiates are drugs that have an effect on our endogenous opiate transmitter system, using endorphins and encephalins. They include heroin, codeine and morphine. All have analgesic effects, and the stronger ones produce euphoria and have a high dependence potential.
Antipsychotics	This group is split into two subgroups. The first is the antischizophrenic drugs, and includes chlorpromazine. The second is the antidepressant drugs. These are further subdivided into the tricyclics, the MAOIs, the SSRIs and the NARIs.
Related topics	Schizophrenia (R1) Mood disorders (R2)

Classes of psychoactive drug

Drugs are classified according to their psychological effects. Given what we have just explored in Topic F2, we need to recognize that:

(i) drugs in the same class can have very different synaptic effects;
(ii) drugs from the same class can affect different neurotransmitter systems; and
(iii) drugs from different classes can affect the same part of the same synapse.

As an example of (i), **amphetamines** and **nicotine** are both stimulants but they affect different parts of the synapse. Amphetamines cause an increase in the release of dopamine from the presynaptic terminal, whereas nicotine has a direct effect on nicotinic receptors. As an example of (ii), **selective serotonin reuptake inhibitors (SSRIs)** and **selective norepinephrine reuptake inhibitors (SNRIs)** are both classes of antidepressants, but work on different neurotransmitter systems. As an example of (iii), both **apomorphine** and **haloperidol** affect the dopamine receptor. However, whilst the former is an agonist and is a stimulant, the latter is an antagonist and is an antipsychotic.

Despite this rather confusing state of affairs, drugs can be largely classified under five main headings. These are stimulants, depressants, hallucinogens, opiates and antipsychotics. Some drugs may belong to a class that is not obvious (e.g. alcohol is a depressant) and some drugs may belong to more than one class depending on the dosage being taken.

Stimulants

Stimulants are drugs that increase neural and behavioral activity. They enhance mood, increase alertness, increase heart rate and blood pressure, and reduce the desire to sleep. A number of these drugs are in common usage. These include **caffeine** and **nicotine**. Other stimulants are used as recreational drugs but may be illegal if used this way. These include **amphetamines** and **cocaine**. However, amphetamines can be prescribed in special circumstances. Each of the stimulants works by slightly different mechanisms. Caffeine works postsynaptically by facilitating second messenger processes and works presynaptically by increasing the entry of calcium ions into the cell (see Topic B3). This, in turn, leads to more neurotransmitter substance being released. It also works as an antagonist at adenosine receptors. Since adenosine inhibits neurotransmitter release, the inhibition of this inhibition results in excitation.

Nicotine is believed to exert its effects by stimulating nicotinic acetylcholine receptors in the brain. Which receptors are affected is not clear, as many of the effects of nicotine can be mediated by its effects on peripheral systems. For example, nicotine increases the release of epinephrine and increases heart rate and blood pressure, but it also activates the parasympathetic nervous system. It is possibly this contrast of effects that gives nicotine the ability to stimulate or relax a person depending on their current mood.

Amphetamine and cocaine are both used as antidotes to fatigue. Amphetamine works by increasing both the leakage into the synaptic cleft and normal release of catecholamines, and also inhibits their reuptake from the synaptic cleft. Behaviorally its effects include increased heart rate, breathing, blood pressure and body temperature. Psychologically, it increases alertness, talkativeness and aggression. However, it can also cause paranoia. Tolerance and dependence are particular problems with amphetamine use (see Topic F4).

Cocaine comes from the coca plant and has been used by South American Indians for centuries. Like amphetamine, it prevents the reuptake of catecholamines from the synaptic cleft. However, it is far more potent than amphetamine. Whilst cocaine is now an illegal drug in most western countries, it was used as an ingredient in Coca Cola between 1886 and 1901. The experiences are very similar to those of amphetamine but the effects last only 20–40 minutes (rather than the hours with amphetamine). This gives cocaine a much higher dependence potential (see Topic F4).

Depressants

Depressants are drugs that cause a general decrease in neural and behavioral activity. They include the minor tranquilizers, like barbiturates and benzodiazepines, and alcohol. All have, as one of their actions, modulation of the GABA receptor. The inclusion of alcohol as a depressant may seem counterintuitive, as most of us use it as though it were a stimulant. However, as we shall see, alcohol has a biphasic action and both phases are a consequence of depressant action.

The **barbiturates** can be used as anesthetics at high enough doses (e.g. pento-barbitol sodium). At lower doses, they produce relaxation and mild euphoria. Like many depressants, they work by increasing the binding of GABA to its receptors. Since GABA is an inhibitory neurotransmitter in many parts of the brain, such enhancement of its action increases the depression in the brain. Barbiturates are not often the drug of choice because they have a number of unwanted side effects. These include insomnia, tremors and digestive problems. They are also very open to abuse and are frequently used for successful suicide because the LD50 is only 2–3 times the ED50. However, barbiturates like pheno-barbitol are still used to treat some forms of epilepsy.

A newer group of depressants are the **benzodiazepines**. These include well-known examples such as diazepam (valium), lorazepam (ativan), alprazolam (xanex), chlordiazepoxide (librium) and temazepam. These are all used to treat symptoms like anxiety, insomnia and stress. They are preferred over barbitu-rates because they have fewer side effects and tend not to be lethal even in high doses. In addition, they have a low abuse potential because they do not lead to the development of physical dependence. However, psychological dependence can be a problem.

Alcohol is possibly the most used psychoactive drug. As mentioned earlier, alcohol is not thought of as a depressant. It has the initial effect of producing euphoria because alcohol depresses inhibitory systems before it depresses excitatory systems. The time course of these effects depends on the quantity ingested and other factors like tolerance and the amount and recency of eating. Alcohol has a high abuse potential and there is some evidence that alcohol abuse might be partly genetically determined. The way in which your body metabolizes alcohol can also influence its effects. Some people are intolerant to alcohol because they are deficient in one of the enzymes that break it down.

Psychedelics

The psychedelics are also referred to as the hallucinogens because this is their major effect. They include cannabis (marijuana), LSD, psilocybin mushrooms and MDMA (ecstasy). These drugs are all illegal in many countries as they have a high abuse potential and, apart from marijuana, have no medical uses. Nevertheless, they represent a very popular source of recreational drug.

Cannabis (marijuana) is a widely used recreational drug. Its mode of action was not known until quite recently when a similarly shaped chemical (anan-damide) was found to be a natural neurotransmitter substance in the brain. The active ingredient in cannabis is **tetrahydrocannabinol** (**THC**), and this attaches to anandamide receptors and mimics its effects. The effects of cannabis include a sense of euphoria, and changes in mood, perception, memory and fine motor skills (Hall et al., 1994). Cannabis is known to have beneficial effects for suffer-ers of glaucoma and multiple sclerosis. However, it also has negative effects such as nausea and anxiety (Williamson and Evans, 2000), and can also lead to deficits in short-term memory (Wilson et al., 1994).

The way in which **LSD** (**lysergic acid diethylamide**) works is complex. It has a similar shape to serotonin and appears to act at serotonergic receptors. However, it seems that the variety of effects that LSD produces is a conse-quence of agonist or antagonist actions at these receptor sites. It is likely that the hallucinogenic properties of LSD are a result of agonist activity at serotoner-gic receptors combined with an action that blocks reuptake of serotonin into the presynaptic cell. Like the other psychedelics, LSD has a number of side effects (Cohen, 1960). It increases energy, lifts mood and increases awareness as well as producing visual hallucinations. Its negative effects include anxiety, nausea and confusion, and it can even induce paranoia at higher doses.

Psilocybin mushrooms (sometimes referred to as 'magic mushrooms') contain the active ingredient **psilocybin**. The actions are very similar to LSD but its effects tend not to last as long. Like LSD, psilocybin shows virtually no physical or psychological dependence.

MDMA (ecstasy) is a drug that has had a recent rise in popularity as a result of the 'rave' culture, even though it was first synthesized in 1912. It is referred to as a 'psychedelic amphetamine' and this describes its actions very well. As such, it is not clear whether this drug belongs more in this class or in the class of stimulants. At low doses, MDMA does not produce hallucinations (Cami et al., 2000) but at higher doses it does (Krystal et al., 1992). MDMA blocks the reuptake of serotonin and also enters the serotonergic cells to cause the leakage of stored serotonin. Its effects are slightly different to other psychedelics as it has a tendency to produce emotional openness and pro-social behavior (Peroutka et al., 1988) in addition to effects like euphoria and increased energy. Negative effects can include inappropriate bonding and difficulty in reaching orgasm. There is a small risk of death but it is not clear whether this might be due to impurity in the drug as sold on the street. Some research has suggested that ecstasy is being sold with as little as 25% MDMA in the pill. MDMA has the potential to produce psychological dependence but many users report that the drug loses its 'magic' after a large number of doses.

Opiates

Opiates are drugs that are derived from the opium poppy. There are naturally occurring opiates like morphine and **codeine**, synthetic opiates like methadone, and partially synthetic opiates like heroin. Heroin is partially synthetic because it is synthesized from morphine.

The opiates exert their effect by mimicking naturally occurring neurotransmitters called **endorphins** and **encephalins**. Endorphins and encephalins are involved in pain perception and have an analgesic effect. They act on opiate receptors that are found in a number of regions of the brain. **Morphine** is the strongest analgesic drug available to us and mimics encephalin action. Tolerance to morphine does develop. However, different opioids work at different types of opiate receptor and so patients who need chronic pain control can be switched from morphine to an alternative if necessary. **Codeine** is a strong analgesic, available without a prescription in many countries. Even so, it is a relatively weak opiate agonist because it does not easily pass through the blood–brain barrier. Both codeine and morphine have a much lower potency than heroin and so their main effect is analgesia rather than euphoria. In addition, even the low potency of codeine produces an antitussive (cough suppressant) action and so it is used in some cough medicines.

Heroin is a very potent opiate and produces a high level of dependence. It is about twice as potent as morphine and, because of this, the effect of taking it is a higher level of euphoria in addition to analgesia. The higher potency is due to the fact that heroin crosses the blood–brain barrier more effectively than morphine. Once in the brain, heroin is converted to morphine but the greater dose–response yields a greater behavioral effect.

Methadone is predominantly used in the treatment of heroin dependence but can be injected subcutaneously for an analgesic effect. As an analgesic, methadone's potency is about the same as morphine, but for preventing the withdrawal symptoms of heroin cessation methadone is about twice as potent as heroin. Methadone works by stopping the 'craving' without inducing all of the psychoactive effects. Furthermore, the effects of methadone last considerably longer than those of heroin.

Antipsychotics   The antipsychotics are really two classes in one. One type of antipsychotic includes drugs that are used to treat schizophrenia (see Topic R1) whilst drugs of the other type are used to treat the affective disorders (see Topic R2). This type includes the antidepressants and the antimanics. Furthermore, there are now a number of different types of antidepressant drug.

The antischizophrenic drugs (also referred to as neuroleptics) include several subgroups. These include the **phenothiazines**, such as chlorpromazine, and the **butyrophenones**, such as haloperidol. Both of these drugs work by blocking the actions of dopamine. Their success in treating the main symptoms of schizophrenia led to the suggestion that this illness is caused by the overactivity of dopamine in parts of the brain. Whilst this is still a favored hypothesis, it cannot account for all cases of schizophrenia, as not all schizophrenics respond to this drug treatment (see Topic R1). Drugs like the ones already mentioned have been problematic because they have many associated side effects. In addition, they are not suitable for all schizophrenics. A recent drug of choice in the UK is clozapine. This works for a wider range of schizophrenics and has far fewer side effects. Furthermore, it is more effective than the other neuroleptics at reducing both the positive and negative symptoms of schizophrenia.

Until recently, there were only two main classes of antidepressant drug. These were the **tricyclic antidepressants**, such as imipramine, and the **monoamine oxidase inhibitors** (MAOIs), such as iproniazid. Both types of drug act to increase the activity of catecholamine neurotransmitters by blocking the reuptake of norepinephrine and serotonin or by preventing their breakdown. The tricyclics are favored because they have a greater effectiveness and have fewer side effects. In fact, the side effects of iproniazid are so severe that it has been withdrawn from use.

A new generation of antidepressants has emerged in recent years. This came about after some research pointed to the possibility that some depressed patients may have a selective lack of serotonin activity without a lack of norepinephrine activity. Drugs like fluoxetine (Prozac) are **selective serotonin reuptake inhibitors** (**SSRIs**) and have had a marked success rate with some people. In addition, the side effects appear to be more tolerable than those associated with either the tricyclics or the MAOIs. A still newer group of antidepressant drugs are the **selective norepinephrine reuptake inhibitors**, so-called because they have a greater specificity for inhibiting norepinephrine reuptake. These drugs are very new and it is too early to make any claims about their success.

When a person suffers depression on its own, the condition is called unipolar depression, or major depressive illness. If the person suffers from periods of mania that alternate with the periods of depression then the condition is called bipolar disorder (see Topic R2). The main drug of choice for treating the mania is **lithium**. However, our understanding of how lithium works is very poor. It may act to increase the rate of reuptake of catecholamines.

F4 TOLERANCE AND DEPENDENCE

Key Notes

Tolerance	This is a dose-related feature of some drugs. The mechanisms of tolerance are described as pharmacodynamic tolerance. Context-specific tolerance is a decrease in drug effect only when the drug is taken in the usual context. The build-up of tolerance means that a higher dose of a drug is needed to achieve the same psychoactive effect. Tolerance to different drug effects builds up at different rates.
Drug dependence and withdrawal	This is a need to continue taking a drug. Dependence can be physical or psychological. The term 'dependence' has replaced the term 'addiction' to reflect the different kinds of dependence and also to reduce the stigma attached to the label. If a drug shows dependence then stopping taking the drug will be associated with negative experiences. These are the withdrawal symptoms and they can be physical and/or psychological.
Related topic	Psychoactive drugs (F3)

Tolerance

Some drugs show a dose-related feature that is dependent upon usage. When certain drugs are used repeatedly the effect of the same dose can be reduced. This is the phenomenon of **tolerance**. The consequence of tolerance is that the user has to keep increasing the dose in order to get the same level of effect. The build-up of tolerance can pose a number of problems for the user. Most drugs have more than one effect. Unwanted effects, which may be dangerous, are called side effects. Tolerance to different effects of a drug builds up at different rates. As a result, drugs that are non-toxic at lower doses might be toxic at higher doses. Furthermore, drugs that have few side effects at low doses may have more at higher doses. Indeed, one can reach the position where the required effects of the drug are decreasing whilst the undesired side effects are increasing, as is the case for the old antipsychotics like chlorpromazine (Petursson and Lader, 1981). A further problem might be cross-tolerance, whereby tolerance to one drug can lead to tolerance to another drug. For example tolerance to the sleep-inducing properties of alcohol can lead to a lessened response to the sleep-inducing properties of low doses of barbiturate.

The main mechanisms by which tolerance occurs are together described as **pharmacodynamic tolerance**. This results from many processes, including an increase in the rate at which the liver metabolizes the drug, where the duration of the effect of the drug will be shortened even though the peak intensity of the effect may remain unchanged. A second sort of pharmacodynamic tolerance results from changes that reduce the response of receptor sites to the drug. In addition, **context-specific tolerance** (also called **behavioral** or **contingent tolerance**) occurs when the effect of a drug decreases when the drug is taken in the same context, although when taken in a different context the original effect returns.

Drug dependence and withdrawal

Drug dependence can be defined as the experiencing of uncontrollable and unpleasant mood states leading to the compulsive use of a drug, despite the possible adverse consequences. In other words, a person has become dependent on a drug when not getting the drug causes real discomfort. This type of dependence usually has both physical and psychological components.

Physical dependence occurs when the cessation of the drug leads to withdrawal symptoms (described below). Virtually all drugs that show dependence show some degree of physical dependence but there are some (e.g. LSD) where no physical dependence occurs. Even the hallucinogens (except LSD) and stimulants, which tend to produce mainly psychological dependence, do yield some degree of physical dependence.

Psychological dependence, on the other hand, develops because taking the drug produces pleasure or alleviates discomfort. In other words, the desire to take the drug becomes habitual. The **positive incentive theory** of drug dependence holds that dependence is not maintained by the avoidance of withdrawal symptoms, but instead by the positive psychological effects that drugs of dependence have. It is thought that psychological dependence may develop because such drugs have an activating effect on the reward sites in the brain (see Topic P2).

A phenomenon called **secondary psychological dependence** can develop as a result of physical dependence. Here, the fear of re-experiencing the physical withdrawal symptoms leads to a psychological dependence.

Some drugs continue to be abused even when there are no longer pleasurable effects and/or when the negative side effects far outweigh the pleasure. Robinson and Berridge (1993) suggested an incentive-sensitization theory to explain this. Their claim is that there are motivational influences on the stimuli associated with taking the drug. These motivational influences become sensitized and this perpetuates the taking of the drug.

It should be noted that dependence in itself is not necessarily detrimental. Dependence is only detrimental if it leads to maladaptive behaviors. So, for example, dependence on caffeine is unlikely to lead to maladaptive behaviors whereas dependence on heroin frequently does.

Withdrawal symptoms are the effects that ensue when a person stops taking a drug. Mostly, the withdrawal effects are the opposite of the effects of taking the drug. So, for example, one of the effects of amphetamine is increased energy and one of the withdrawal symptoms is fatigue. Withdrawal symptoms occur when tolerance has been built up to a drug.

You may have noticed that we have not yet used the term **addiction**. Addiction and dependence are synonymous, but many researchers are trying to reduce the use of the term addiction. The term disguises the distinction between physical and psychological dependence, and it has negative connotations. To state that a person is an addict gives the person the stigma of a negative label, but to state that a person has a physical dependence is less stigmatized. Addiction implies illegality. However, not all drugs of dependence (addictive drugs) are illegal. Alcohol, which is legally available in most western countries, has a very high dependence potential, and is enormously costly in personal and health terms.

G1 THE VISUAL SYSTEM

Key Notes

The primary visual pathway	The visual pathway starts at the eye and ends at the visual cortex. Along the way, some axons cross the midline from one side to the other and others do not cross. Foveal information travels to both hemispheres.
The eye	The eye has a complex structure. Its purpose is to focus light onto the photoreceptors. To this end, the important components are the iris, the lens and the retina. There is a blind spot where the optic nerve leaves the retina. Owing to top-down processes this is not normally visible.
Photoreceptors	Photoreceptors are the sensory cells that convert the incoming light into neuronal signals. Rods can respond to low light levels whereas cones require normal light levels. There are three types of cone which respond to different wavelengths of light. We cannot see color in low light when only rods are active.
Retinal ganglion cells	Processing of light information starts in the retinal ganglion cells. Information from the photoreceptors is passed to the retinal ganglion cells but surrounding cells can affect the precise pattern of this input.
The lateral geniculate nucleus	The lateral geniculate nucleus is the first place where light from the two eyes is combined. It is comprised of six layers and the way in which information enters the layers is highly organized.
The visual cortex	The visual cortex is where most of the primary processing for object recognition occurs. The cortex consists of a primary visual area and secondary areas.
Related topic	Coding in the retina and lateral geniculate nucleus (G2)

The primary visual pathway

The visual pathway starts at the eye with the **photoreceptors**. These send an input to cells called **retinal ganglion cells** in the **retina**, at the back of the eye. From here, the pathway goes to the **lateral geniculate nucleus** and then to the **visual cortex**. The pathway, for an object positioned on the left of center when we are looking straight ahead, is illustrated in *Fig. 1*.

The most important thing to note about the figure is the way in which light travels from the two eyes to the visual cortex. If we trace light that comes in from the left of center (the left visual field) when we are looking straight ahead, we see that the light hits the inside (nasal) part of the left retina and the outside (lateral) part of the right retina. If we now follow the nasal pathway, we see that it crosses the midline at the **optic chiasma**. From here, the information

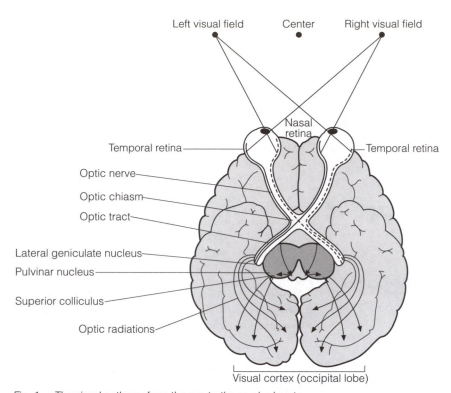

Fig. 1.   The visual pathway from the eye to the cerebral cortex.

travels to the right lateral geniculate nucleus and on to the right visual cortex. The path of the nasal visual field is described as **contralateral**. If we now follow the lateral pathway, we see that it does not cross the midline at the optic chiasma but continues on the same side to the right lateral geniculate nucleus. From here, the information travels on to the right visual cortex. This path is described as **ipsilateral**. The net result is that information from the left of center when we are looking straight ahead (the left visual field), all goes only to the right hemisphere, even though the light goes to both eyes.

The same principle is true for information from the right of center (the right visual field), only all of the information goes to the left hemisphere. The exception to this is that information from objects we are directly focusing on (called foveal vision) goes to both hemispheres.

The eye   *Fig. 2* shows a simplified diagram of the eye. It is a complex structure with two fluid-filled chambers. These fluids mostly serve to maintain the shape of the eye. Between these two fluid chambers is a **lens**, which is partially covered by a circular band of muscle called the **iris**. The muscles of the iris expand and contract to alter the size of the hole that lets light into the eye (the **pupil**). The lens is capable of changing its shape slightly so that light from near or far objects can be focused onto the retina. The **retina** is the layer at the back of the eye that contains the **photoreceptors**. Its structure is shown in *Fig. 3*. Note that the light enters the eye and has to pass through a number of cell layers before reaching the photoreceptors. The part of the retina where the object we are

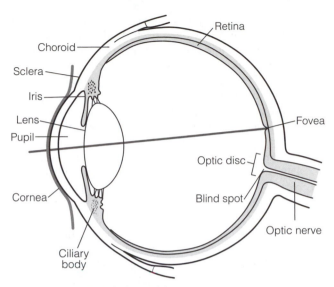

Fig. 2.   A simplified diagram of the eye.

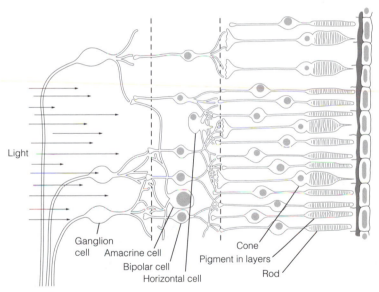

Fig. 3.   The retina.

attending to is focused is called the **fovea**. There is also a **blind spot** on the retina. This is where the **optic nerve** leaves the eye. As this region is densely packed with neurons, there is no room for photoreceptors. Hence, this region of the retina is literally blind.

That we do not see a blind hole in our vision where the optic nerve leaves is due to the **top-down processing** of visual information. Our perception is not simply the result of the information that enters the eye. We have an expectation about what we will see when we look at anything. This expectation is based on what makes the best sense in your current context. If, for example, you walk out of your front door it probably makes sense for you to expect to see a road, people, and so on. So your perceptual system has readied itself for this sort of visual input. This process is a top-down process and the act of seeing is partly a confirmation of your expectations. In the same way, we can fill in the gap in our visual field with what we expect to be there. So for example, if your blind spot relates to a patch of repeating wallpaper, your brain will fill in the appropriate pattern. If there happens to be a tear in the wallpaper at that point, you will fail to see it until it presents itself to a part of your visual field not covered by the blind spot.

Photoreceptors

The photoreceptors are the cells that convert light information into electrical signals. There are two types of photoreceptor in the human eye, the rods and the cones. The **rods** are used for seeing in low light conditions but are not capable of encoding color information. The **cones**, by contrast, are used for color vision but only work when the light level is higher. Hence, we are not able to see in color at night. There are three different kinds of cone that respond to different ranges of wavelengths of light (*Fig. 4*). Roughly, these wavelengths correspond to red, green or blue light. At the fovea, described above, there are no rods, only cones. In the periphery, rods by far outnumber the cones.

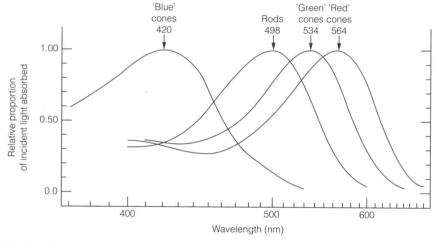

Fig. 4.   Response ranges of the rods and cones.

Retinal ganglion cells

The photoreceptors activate **bipolar cells**, which then activate **retinal ganglion cells**. Retinal ganglion cells are also activated by **amacrine cells**. The role of amacrine cells is not really known. However, the net result of all of this activity is that the ganglion cells receive enough information to start to process the visual input for its meaning. The retinal ganglion cells give rise to the nerve fibers that leave the retina via the optic nerve.

There are two types of ganglion cell. **P ganglion cells** seem to respond to color and are used for the recognition of fine detail, whereas **M cells** are better at responding to large objects or to movement. The receptive fields (described later) of these cells also differ.

The lateral geniculate nucleus

The **lateral geniculate nucleus** (**LGN**) receives input from both eyes. It is a six-layered nucleus and the input to each of the layers is organized by eye. Layers 2, 3 and 5 receive ipsilateral input whereas layers 1, 4 and 6 receive contralateral input. Furthermore, the layers of the LGN receive inputs from only one or other of the two types of retinal ganglion cell. Layers 1 and 2 receive input only from M ganglion cells and so are responsive to movement and large objects. This is the **magnocellular system** (hence the label 'M cells'). Layers 3–6 receive information from the P ganglion cells and so are responsive to color and fine detail. This is the **parvocellular system**.

The visual cortex Information comes to the visual cortex from the LGN via the optic radiation (see *Fig. 1*). Like the LGN, the visual cortex has six layers. In addition, the visual cortex has a number of distinct areas. The information from the LGN goes first to the primary visual cortex (area V1), which is also called the **striate cortex** because of its striped appearance. There are also a number of **extrastriate** areas, the main ones being areas V2, V3, V4 and V5 (also called area MT — middle temporal). We will concentrate in this section on area V1.

Information from the LGN goes to layer 4 of the primary visual cortex. This layer is itself divided into sublayers, and the magnocellular and parvocellular LGN cells project to different sublayers. A separate sublayer of layer 4 contains **simple** and layers 2, 3 and 6 contain **complex cells** (see Topic G3). In addition to this organization, area V1 is further organized according to ocular dominance. As you move horizontally over the surface of the cortex you find alternating regions that have a preference for information from either the left or right eye (although they all do respond to both eyes). *Fig. 5* shows that as you move across the surface of the cortex in a perpendicular direction, the cortex is organized in columns of cells that all respond to stimuli in a particular orientation. That orientation gradually changes as you move across the surface.

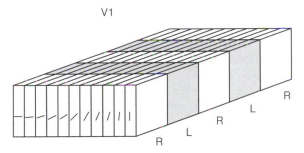

Fig. 5.   The columnar organization of the visual cortex.

G2 CODING IN THE RETINA AND LATERAL GENICULATE NUCLEUS

Key Notes

Activity in the photoreceptors	There are two types of photoreceptor: the rods and the cones. Rods are used for low light, black and white vision. Bleaching of a pigment, rhodopsin, in rods leads to a reduced release of glutamate, which signals the presence of light. Cones are only capable of operating in good light. There are different opsins in three different cones types, responding to different wavelength ranges, and this allows us to see colors.
The role of the retinal ganglion cells	The retinal ganglion cells start to process the light signals for meaning. Each cell has a receptive field with opposed on-off centers and surrounds. X cells respond to stationary points of light. Y cells respond to changes or moving stimuli. W cells have more complicated receptive fields. Their centers respond to either light or dark.
The role of the lateral geniculate nucleus	The lateral geniculate nucleus is a highly topographically-organized relay station. It has receptive fields similar to those of the retinal ganglion cells. Color coding in the form of opponent processes is evident in the LGN.
Related topic	The visual system (G1)

Activity in the photoreceptors

The way in which the rods and cones convert light into electrical signals is by way of complex chemical processes. Rods contain a pigment called **rhodopsin** that becomes bleached when exposed to light. This bleaching leads to a cascade of events that result in a reduced release of **glutamate** (a neurotransmitter) by the rod cell. This is the signal for the presence of light. Note that this signal is transmitted through inhibition rather than excitation. The process in the cones is very similar, except that the **opsins** in each cone are different and this probably accounts for the different wavelengths of light responded to.

As well as being activated by light, the rods and cones can activate each other. One way in which this can occur is via activation of the horizontal cells (see *Fig. 3*, Topic G1). This system serves to reduce the responses of neighboring cells exposed to diffuse stimulation and to heighten the responses of cells at places where the light level changes (e.g. borders and outlines).

The role of the retinal ganglion cells

The retinal ganglion cells are the first place in which the visual stimulus is analyzed for form. Our understanding of the responses by retinal ganglion cells is due to the work of Hubel and Wiesel in the 1960s. They (see Hubel and Wiesel, 1979) discovered that each retinal ganglion cell responds to a pattern of light in a particular place in the visual field. The area in space that a cell responds to is called its **receptive field**. For retinal ganglion cells, the fields look like the ones shown in *Fig. 1*.

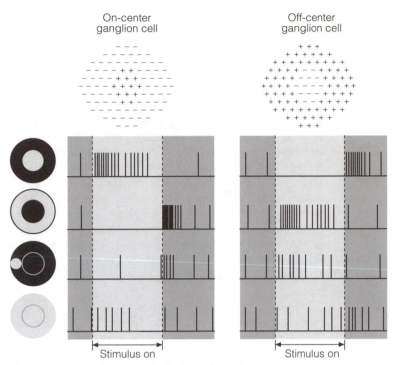

Fig. 1.   The receptive fields of retinal ganglion cells.

For an **on-center field**, the retinal ganglion cell maximally increases its firing to a stimulus that is a circle of light that completely covers the on center and does not at all cover the ring surround. A ring of light that completely covers the surround and does not at all cover the inner circle will maximally turn the cell off. Note that when the same level of light covers the entire field, there is still a **resting**, or **tonic** firing rate in the retinal ganglion cell. This is a common feature of neurons as it allows for the neuron to register both increases and decreases in level of stimulation. For **off-center** receptive fields the firing patterns are exactly the reverse of on-center cells.

More recently, three types of on-center and off-center fields have been discovered. These have been labeled X cells, Y cells and W cells. **X cells** seem to respond to stationary points of light and can distinguish fine grain detail. **Y cells**, on the other hand, respond best to changes in illumination or moving stimuli. **W cells** have more complicated receptive fields than simple on-center or off-center fields. Their field centers respond to either light or dark rather than to just one or the other.

The role of the lateral geniculate nucleus

The lateral geniculate nucleus (LGN) serves mainly as an organizational relay station. The cells are topographically organized as described in Topic G1. It is noteworthy that the LGN is the first place for which the receptive fields all relate to the opposite side of the body. The fields, themselves, are very similar to those of the retinal ganglion cells. LGN receptive fields are also made up of a circle surrounded by a ring and which is either on-center or off-center. In addition, both X cells and Y cells are also seen in the LGN. There appears to be some relationship between X and Y cells, and P-cell layers and M-cell layers. The cells of the four P-cell layers have X-cell properties and respond to fine grain detail. They also respond to lights of different colors. The cells of the two M-cell layers have Y-cell properties in that they respond most to moving stimuli.

Color discrimination also occurs in the LGN. The first theory of color vision was proposed by Thomas Young and Hermann von Helmholtz. Their theory (the Young–Helmholtz Theory or Trichromatic Colour Theory) suggested that in order to make any color people only need three color receptors: red, green and blue. One remarkable feature of their theory is that they were working around 1800 and it wasn't until the late 1970s that we discovered the wavelengths of light that the cones respond to. However, the Young–Helmholtz theory could not account for negative afterimages. An example of a negative afterimage is that if you stare at a green square and then look at a white square, the white square appears red. The opponent-process theory proposed by Hering in 1878 solved this dilemma by suggesting that coding for color occurs as three opponent processes. Hering suggested that red and green, blue and yellow, and black and white were all opposing colors and that individual neurons code for a pair.

Evidence for this opponent process has been found in the LGN. There are two types of opponent cells in the LGN. These show red–green and yellow–blue opposition. Red–green opponent cells register either red *and not green* or green *and not red*. In other words, saying that we see green negates our ability to simultaneously say we see red, and vice versa. The blue–yellow opponent cells work in a similar fashion. Yellow is an opponent color because red and green cones firing together give us the perception of yellow.

G3 CODING IN THE VISUAL CORTEX

Key Notes

Coding for form, orientation and movement

Hubel and Weisel identified a number of different kinds of cell in the visual cortex. The *simple* cells in layer 4 of the visual cortex code for lines or edges. *Complex* cells in layers 2, 3 and 6 are selective to contours in particular orientations, and respond to movement. *Hypercomplex* cells respond to corners or line ends.

Coding for stereoscopic depth

Retinal disparity is an important mechanism for humans and relies on stereoscopic vision. Cells in layer 4 of the visual cortex are monocular, but cells in other layers receive binocular inputs.

Coding for color

The cones provide basic information about color, but some phenomena, like color constancy, require complex processing. The opponent-color theory helps explain this. Cells called *blobs* and *interblobs* in layers 2 and 3 of area V1 code for color. Like LGN cells they are opponent-color cells. The global perception occurs in area V4.

Coding for movement

The detection of movement is critical for tracking objects across the visual environment. Some motion detection occurs in the primary visual cortex but most of it occurs in area V5.

Related topics

The visual system (G1)

Coding in the retina and lateral geniculate nucleus (G2)

Coding for form, orientation and movement

The cells of layer 4 of visual area V1 were labeled **simple cells** by Hubel and Wiesel. This label refers to the fact that the receptive fields of these cells all code for simple visual features such as lines and edges. The types of receptive field of these cells are shown in *Fig. 1*. Note that they can have an on-center or an off-center but this does not have to be central within the field. Alternatively, the field can be split in half so that it registers an edge. *Fig. 2* illustrates how these fields are constructed from the concentric fields of the retinal ganglion cells. As well as responding maximally to a line or an edge, simple cells have a preference for lines and edges of a particular orientation. *Fig. 3* illustrates a cell that responds maximally to a horizontal line of light. There are, however, some common properties of all simple cells. All respond best to a stimulus of a particular orientation and a stimulus positioned so that it borders the on- and off-zones. Furthermore, the on- and off-zones of all simple cells exactly cancel each other out.

Layers 2, 3 and 6 of visual area V1 contain **complex cells**. These cells receive input from the simple cells. They are similar to simple cells in that they require the specific orientation of a light/dark boundary. They are also similar in that light over the whole receptive field leads to no overall response. However, the exact

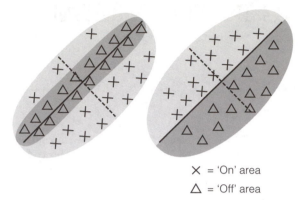

X = 'On' area

△ = 'Off' area

Fig. 1. Line and edge receptive fields of the visual cortex.

Circular on-center
retinal ganglion fields

Cortical simple
cell line detector

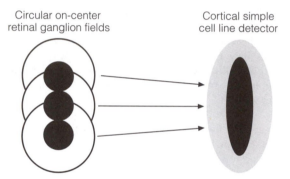

Fig. 2. The construction of a line receptive field from circular receptive fields.

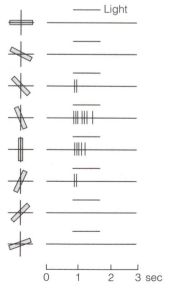

Fig. 3. Diagram showing preferred line orientation of a cortical receptive field.

position of the light/dark boundary within the receptive field is not important. As long as the boundary falls somewhere within the receptive field, the cell will alter its firing rate. These cells also respond to boundaries that are moving across the field. A number of different classes of complex cell have been identified.

Another type of cell has been identified. This type of cell responds best to a line that stops within the receptive field or to a corner placed within the field. These cells were previously referred to as **hypercomplex cells**. They are now labeled as further varieties of complex cell.

All of these varieties of cortical cell suggest that the visual system builds up a pattern of lines and edges from the 'dots' of the retinal ganglion cells. This strongly points to a bottom-up approach to visual processing. However, it would be premature to jump to such a conclusion, as there is a lot of evidence that does not fit with a purely bottom-up explanation.

Coding for stereoscopic depth

The most important form of depth information in humans comes from **retinal disparity**. This is the fact that the images from two objects that are at different distances do not fall on the same parts of the retinas of the two eyes. We can use the degree of disparity as a relative marker of the distance between the two objects. The question that many have asked is how we can match the two images to be able to measure disparity. It appears that there may be some cells within the visual cortex that are especially for this function. This is supported by the fact that in layer 4 of the cortex (the input layer) the cells are monocular. However, above and below this layer we find that the simple and complex cells are binocular. Further support comes from the developmental disorder of **strabismus**, in which the binocular cells in the visual cortex are not simultaneously activated by the same stimulus. The result is that these cells become monocular and the child permanently loses **stereopsis**.

There is other evidence that there are three types of disparity detector in the visual cortex. One type of cell is active when the images are displaced outward (**close neurons**), a second type is active when the images are displaced inwards (**far neurons**), and the third type are active when there is no retinal disparity (**in focus neurons**).

Coding for color

Coding for color not only occurs through the differential responding of the three types of cone and by the opponent processes in the lateral geniculate nucleus. It also occurs in the visual cortex. In layers 2 and 3 of area V1 there are cells called **blobs** and **interblobs** that code for color rather than for orientation. They receive their input from the **parvocellular layers** of the lateral geniculate nucleus.

Blobs are also opponent-color cells that are either red–green or blue–yellow (as described for the LGN cells in Topic G2). Even so, color perception is a global perception of the whole visual scene. The responses of these wavelength-specific cells do not give us the perceived color. This global perception occurs in area V4. Here, cells only respond to one narrow band of wavelength. Indeed, damage to V4 can totally eliminate the ability to perceive color. From V4, information is sent to the temporal lobe for further color processing. Presumably this is the start of the integration of color information with other types of information – memory, language, and so on.

Coding for movement

As already mentioned, some detection of motion is processed by the complex cells of area V1. However, the main processing of motion information occurs in the cells of area V5 (also called area MT). The cells in area V5 are organized such that the cells in a column respond to movement in a particular direction. These cells respond to the direction of movement irrespective of the orientation of the object. It is believed that motion perception involves opponent mechanisms similar to those involved in color perception. However, the details of this opponency are not fully understood.

G4 Two visual systems

Key Notes

Why two distinct pathways?	Visual stimuli provide two distinct types of information. One type concerns what the objects being seen are and the other type concerns where they are.
The dorsal stream	The dorsal stream is the 'where' stream. It originates in the M retinal ganglion cells, then passes from the magnocellular cells of the lateral geniculate nucleus to area VI of the visual cortex.
The ventral stream	The ventral stream is the 'what' stream. It originates in the P retinal ganglion cells, via the parvocellular cells of the LGN, which project to areas 2 and 3 of visual area V1. From here, information is sent to the blobs and interblobs of this area.
Related topic	The visual system (G1)

Why two distinct pathways?

As well as visual stimuli giving us information about form, color, depth and motion, we are able to distinguish visual stimuli along a different dichotomy. That dichotomy is the distinction between what an object is and where in space that object is. These different types of information are processed by two distinct pathways in the visual system.

The dorsal stream

The **dorsal stream** is also called the 'where' stream because it deals mainly with information about *where* an object is. Initially, the dorsal stream was believed to be a visual pathway that involved the **superior colliculus** rather than the visual cortex. However, later work by Ungerleider and Mishkin (1982) pointed to the suggestion that both *where* and *what* information are processed by different parts of the visual cortex. Indeed, both streams are found to extend beyond the visual cortex.

The dorsal stream actually starts in the magnocellular cells of the lateral geniculate nucleus (LGN). Fibers from these cells project to layer 4 of area V1 and then project to area V2. Dynamic form information is processed by area V3, whereas motion information is processed by area V5. Information in the dorsal stream leaves the visual cortex and goes to the parietal cortex. There is also a projection to areas of the frontal cortex that control movement.

It is the combination of dynamic form and motion information that allows us to determine where an object is. The need to monitor where information is closely related to the ability to make visually guided movements. Capabilities like reaching and grasping critically require information about where an object is.

The ventral stream

The **ventral stream** is also called the 'what' stream because it deals mainly with information about *what* an object is. Unlike the controversy surrounding the dorsal stream, all of the early data concerning the ventral stream pointed to the suggestion that the processing of *what* information was cortical in origin. The anatomy of the ventral stream starts with the parvocellular cells of the LGN. These project to areas 2 and 3 of visual area V1. From here, information is sent to the blobs and interblobs of this area. Information about form then goes to area V2 and on to area V4. Information about color goes to area V3 and then to area V4. From area V4, information is sent to the temporal cortex, presumably to allow the 'what' information to be integrated with other information processing.

There has been a suggestion that cells of the temporal cortex ought to fire in response to a whole object. This has been characterized by the idea of a '**grandmother cell**' (one that fires only when you look at your grandmother). Whilst no such cells have been found, it is likely that the cells of the temporal cortex code for the building blocks of whole objects. What is not clear is whether this template approach can support the diversity of form recognition that we are capable of.

H1 THE AUDITORY SYSTEM

Key Notes

The structure of the ear	The ear can be divided into three parts. The external ear consists of a pinna and an external auditory meatus, and channels sound to the tympanic membrane. The middle ear amplifies the sound, through three tiny bones, the malleus, incus and stapes, passing it from the tympanum to the oval window. The inner ear contains the cochlea, which changes sound waves to auditory neural signals. The cochlea is a coiled, fluid-filled structure containing sensory hair cells along the Organ of Corti. These are connected to the auditory nerve.
The auditory pathway	High frequency sounds produce maximal vibration near the base of the basilar membrane, and low frequency near its apex. Hair cells in the cochlea of the inner ear convert sound energy into neuronal signals. The neurons leave via the auditory nerve and travel to the brain stem. From here one branch passes to the superior olive and the other to the inferior colliculus. Outputs from the inferior colliculus travel to the auditory cortex via the medial geniculate nucleus. Binaural information is combined in the superior olivary nuclei. Two different pathways lead from the medial geniculate nucleus to the primary and secondary auditory cortexes.
Deafness	Conduction deafness refers to a loss of transmission from the eardrum to the hairs of the cochlea. Outer ear deafness can result from build-up of wax, from a tumor or from being born without an external auditory meatus. Middle ear deafness can occur if the eardrum is ruptured. Inner ear deafness is the result of damage to the hair cells, which can result from exposure to loud sounds. Nerve deafness refers to hearing loss as a result of damage to the auditory nervous system.
Related topic	Coding for pitch and timbre (H2)

The structure of the ear

The ear has three distinctive regions (*Fig. 1*). The *outer ear* consists of a **pinna** and an **ear canal** (the external auditory meatus). The pinna has a number of infolds that help to direct sound waves towards the ear canal. Not all frequencies are directed equally and there tends to be a favoring of frequencies between 2000 and 5000 Hz (the frequencies within which human speech operates). We will see in Topic H3 that the pinnae are important for the localization of sound. Each pair of pinna is unique to that person. In an experiment in which people were fed sound via moulds made from other people's pinnae, the sounds were very distorted. At the end of the ear canal is the eardrum. This is a sensitive membrane that vibrates when sound waves hit it.

The ear canal leads to the **eardrum** (or **tympanic membrane**) that separates the outer ear from the middle ear. The *middle ear* contains three tiny ossicles (bones), the **malleus**, **incus** and **stapes** (Latin words for hammer, anvil and

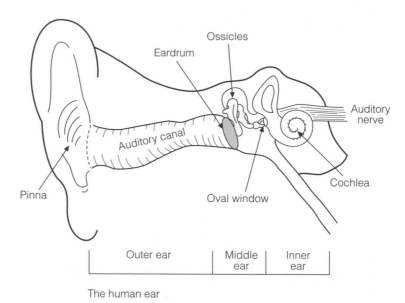

The human ear

Fig. 1. Diagram showing the three regions of the ear.

stirrup, which describe their shapes). These bones are the smallest in the body and their role is to amplify the small variations in pressure detected by the eardrum and feed these amplified vibrations to the inner ear via the oval window. Muscles attached to the malleus and the stapes modify the way in which the bones respond to pressure changes at the eardrum. This acts as a safety mechanism. They stiffen the bones in response to very loud sounds and thereby dampen the effect, thereby minimizing damage to the system. They are also active in response to body movements and to vocalization, so that we are less sensitive to the sounds of our own bodies.

The inner ear is complex, and contains the cochlea, the structure that changes sound waves to auditory neural signals, and the vestibular organs, which are central to the somatosensory system (see Topic I1). The **cochlea** is a fluid-filled structure containing **hair cells**. These hair cells are the sensory cells that create the neural signal that is carried to the brain. The cochlea is a coiled structure that contains three separate canals that run along its length, the **tympanic canal**, the **middle canal** and the vestibular canal (*Fig. 2*). Only the first two of these are of interest to us in our consideration of hearing. The fluid in the canals cannot be compressed and so indentations of the oval window are matched by 'outdentations' of a round window that also separates the inner ear from the middle ear. Between the tympanic and middle canals is a membrane called the **basilar membrane**. This membrane is wider at the apex (which is furthest from the oval window) than at its base and it vibrates in response to changes in the oval window (*Fig. 3*). On top of the basilar membrane is the **organ of Corti**, which consists of two layers of hair cells, a set of supporting cells and the auditory nerve fibers. There are one row of inner hair cells and three rows of outer hair cells. The hair cells have rows of hairs (called **stereocilia** or just **cilia**) on top of them and these cilia are of different lengths so as to present an inclined plane to the moving fluid. The inner hair cells are connected to the **auditory nerve** and are responsible for 90–95% of our hearing. The outer hair cells also

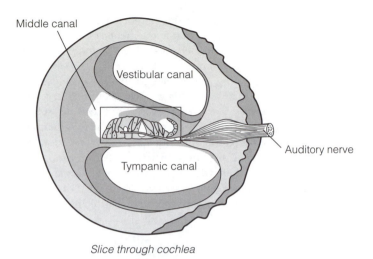

Slice through cochlea

Fig. 2.   The canals of the inner ear.

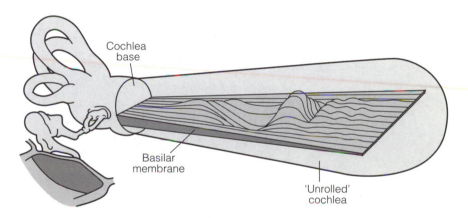

Fig. 3.   The basilar membrane of the cochlea.

connect to the CNS via the auditory nerve but their role seems to be more one of moderating the response of the inner hair cells.

The auditory pathway

Sound waves are converted into neuronal signals by virtue of the vibration properties of the basilar membrane. Movements of the oval window cause pressure changes in the fluid of the inner ear. Experiments in the 1930s conducted by Georg von Bekesy showed that if the sound is a high frequency sound, the result is maximal vibration of the basilar membrane near its base. However, if the sound is a low frequency sound, the maximal vibration is near the apex of the basilar membrane. The angle at which the stereocilia on the inner hair cells are oriented causes them to move in response to vibrations in the basilar membrane. The movements of the hairs are translated into the opening and closing of ion channels in the stereocilia themselves (*Fig. 4*). The ion channels are physically gated so that movements of the stereocilia physically open or close the channels. Furthermore, these channels can be opened and closed at rates of thousands of times per second.

When the channels on the stereocilia open, potassium (K^+) ions flood in and depolarize the hair cell. This allows calcium (Ca^{2+}) ion channels to open and Ca^{2+} ions to flood in. This causes the cell to release its neurotransmitter substance, which for these cells is probably glutamate. The glutamate then stimulates the afferent, auditory nerve neurons. However, the vibrations of the basilar membrane are not sufficiently distinctive to give the level of frequency discrimination that we are capable of. Humans can distinguish between frequencies that are only 2 Hz apart. This level of acuity is achieved by a tuning process that is a function of the outer hair cells. For any given sound input, the

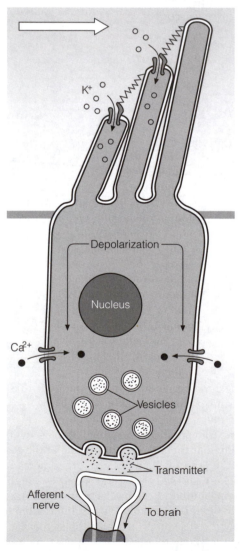

Fig. 4.   The opening of ion channels by the stereocilia.

outer hair cells appear to amplify the movements of certain regions of the basilar membrane and dampen other regions.

Outputs from the ear travel via the **8th cranial (auditory) nerve**. Auditory nerve fibers go from each ear to both sides of the brain. As the auditory nerve enters the brain stem it divides into two pathways. These go separately to the dorsal and ventral **cochlear nuclei**, which are located in the medulla oblongata. Different parts of these nuclei respond to different frequencies of sound. The first places at which information from the two ears (binaural information) are combined are the **superior olivary nuclei**, which have a direct input from the cochlear nuclei. The cochlear nuclei also project to the **inferior colliculus**, which in turn projects to the **medial geniculate nucleus**. Two different pathways lead from the medial geniculate nucleus to the auditory cortex. These go to different parts of the cortex, the **primary auditory cortex** and the **secondary auditory cortex**. One output from the primary auditory cortex goes back to the medial geniculate nucleus and the inferior colliculus, but other outputs go to association cortex and to the speech areas (**Broca's area** and **Wernicke's area**; see Topic Q2).

At each of the nuclei in the auditory pathway, cells are organized in order of frequency. This is referred to as a **tonotopic organization**. Furthermore, cells of the auditory cortex have auditory fields, in a similar way to that described in section G for cells of the visual cortex. These cells may respond maximally to sound sources, to species-specific sounds or to the movement of sound across the cell's field.

Deafness

Deafness can occur either as a result of damage to the mechanics of the ear or due to damage to the nerve fibers that convey the auditory information to the brain or process it within the brain. The former is known as conduction deafness and has three forms; outer ear deafness, middle ear deafness and inner ear deafness. The latter has two forms; nerve deafness and central deafness.

Outer ear deafness can result from an excess build-up of the waxy substance secreted by the walls of the external auditory meatus. This will clog the ear canal and prevent the sound waves from getting through. More permanent deafness can result from a tumor or from being born without an external auditory meatus. **Middle ear deafness** can occur if the eardrum is ruptured. This can be rectified by the implanting of an artificial eardrum. However, aging can lead to stiffness of the ossicles through arthritis or rheumatism. This can impair the quality of the transmission of auditory information. It can usually be rectified by wearing a hearing aid. **Inner ear deafness** is the result of damage to the hair cells. This can occur in people who work with loud noises and unprotected ears (e.g. rock musicians). Alternatively, it can occur as a result of inherited progressive hearing loss in which the hair cells gradually degenerate. In some cases, a cochlear implant can be used to directly stimulate the auditory nerve in response to a microphone placed behind the ear.

Nerve deafness is damage to the auditory nerve as a result of injury or infection or as a result of an inherited, degenerative disorder. There is currently no cure for this type of deafness. **Central deafness** is damage to the more central processing elements of hearing. Again, the sources of damage can be injury or infection and, again, there is currently no cure.

H2 CODING FOR PITCH AND TIMBRE

Key Notes

The physical stimulus	The physical stimulus that gives rise to pitch is frequency. Frequency is a property of sound, whereas pitch is a subjective property of the processing of sound.
Place coding	One way in which we can determine the pitch of the sound is by noting the place of maximal vibration on the basilar membrane. This works for medium to high frequency sounds.
Rate coding	Low frequency sounds are not well coded by place. However, they can be coded by the rate at which the hair cells fire.
Coding for timbre	Timbre is the name given to the characteristic quality of the sound of an instrument. The overtones are initially coded separately but are combined in a complex manner.
Related topic	The auditory system (H1)

The physical stimulus

Sound waves have different frequencies and we perceive these as differences in pitch. However, there is not necessarily a one-to-one relationship between changes in frequency and changes in perceived pitch. As already noted, differences in frequency of the physical stimulus are translated into differences in the vibratory characteristics of the basilar membrane in the cochlea of the inner ear. Nevertheless, high and low pitched sounds are not processed by the same mechanisms.

Place coding

One theory about the way in which we are able to distinguish differences in pitch suggests that the **basilar membrane** vibrates maximally at different places along its length, depending on the pitch of the sound. When a high frequency sound is heard, the thinner base of the basilar membrane vibrates maximally. As the sound lowers in frequency, the maximal vibration moves towards the thicker apex of the membrane. This is very similar to the strings of a guitar, where the thinner strings give higher frequency sounds. Variations in the *loudness* of sounds are transmitted by variations in the rate of firing of auditory neurons.

Evidence that this **place coding theory** is correct comes from two sources. High doses of antibiotics, such as kanamycin, can cause damage to the basilar membrane by inducing hair cell loss. This loss starts at the base of the membrane and higher frequencies are affected first (Stebbins et al., 1969). The other source of evidence is the success of cochlear implants. If a person is deaf because of the loss of basilar membrane stimulation, a **cochlear implant** device can be implanted that mechanically converts the sound waves into vibrations at different regions of the membrane according to the frequency. Provided that the basilar membrane and all later aspects of the auditory system are functioning normally, hearing will be restored.

Rate coding   Sounds at the lowest frequencies we can detect are not distinguished by place coding. Below 200 Hz, no particular region of the basilar membrane cells appears to respond best. The method by which these low frequencies are identified is by the temporal sequence of firing at the apical end of the basilar membrane. In other words, low frequencies are detected by **rate coding**. Since the pitch is determined by the rate of firing, the loudness of low frequency sounds is probably determined by the number of cells firing. It is possible that some form of rate coding also takes place for frequencies above 200 Hz, but this is likely to serve only as a secondary mechanism.

Coding for timbre   We can recognize the difference between a violin and a clarinet, even though they may be playing notes of the same pitch. The sound we hear is made up of a **fundamental frequency** and a number of **overtones** that are all multiples of the fundamental frequency. Different instruments have different overtones and this gives them their different quality characteristics.

Each overtone will maximally excite a different part of the basilar membrane. This pattern of basilar membrane excitation is unique to a particular note of a particular pitch played on a particular instrument. The input pattern is transferred to the auditory association cortex where the sound is interpreted. As the overtones of an instrument are constantly changing, the sound that we hear is a dynamic feature and this adds to our appreciation of the qualities of the instrument.

H3 THE LOCALIZATION OF SOUND

Key Notes

Having two ears	Having two ears allows us to hear binaurally. The differences in arrival of sound at the two ears give us many clues about the location of the sound source.
Phase difference and sonic shadow	Three aids to localizing sound either side of the midline are the time differences at which the sound arrives at each ear (arrival time difference), differences in phase of the sound wave at the two ears (phase difference) and intensity differences at each ear (sonic shadow).
Monaural cues	Even people with one ear can localize sound to some extent. The shape of the pinna reflects the sound in a way that gives an indication of the sound's location (spectral filtering).
Differentiating 'in front' from 'behind'	This is not easy as phase differences and sonic shadow cannot be used. Differentiating in front from behind relies heavily on monaural cues.
Related topic	The auditory system (H1)

Having two ears

In the same way that having two eyes allows for greater visual abilities through stereoscopic vision, so having two ears affords a greater skill in hearing. The use of two ears is called **binaural detection**. Without two ears, our ability to locate a sound source is diminished, although, as we shall see, not completely eradicated.

Phase difference and sonic shadow

When sound comes from one side or the other, two cues are available to help locate its source. The first cue comes from the fact that the sound will reach one ear prior to the other. This creates an **arrival time difference** (also known as a **latency difference**), which can be detected by neurons in the **superior olive**. Cells here are tuned to particular arrival time differences and can detect differences in arrival time at the two ears that are as small as a fraction of a millisecond. Differences in arrival time are adequate to locate a sound source if the sound is a simple and discrete one (for example, a click), but a variation of this is needed to explain our ability to locate more complex or continuous sounds.

    Phase difference refers to the fact that different portions (phases) of the sound wave arrive at each ear at any one time. That is, while one ear is stimulated by a high pressure part of the sound wave, the other might be stimulated by a lower pressure part. These phase differences have the following consequence. Sound waves cause the eardrum to be pushed in and pulled out (i.e. to vibrate). Phase differences will mean that whilst, say, the right eardrum is pushed in, the left eardrum might simultaneously be pulled out. If the eardrum positions were exactly opposite, then the phase difference would be 180°. Given that the phase difference for sounds directly in the midline will be zero and the maximal phase difference is 180°, the combination of phase difference and sound frequency will give an indication of the sound's location. This information is more useful for low frequency sounds than for high frequency sounds because of the wavelength and the distance separating the ears. Neurons of the **medial superior olive** are able to use this phase difference as an aid to locating the source of the sound.

A second cue to location comes from sound intensity differences that arise from what is called the **sonic shadow**. This method of detection is particularly useful for high frequency sounds when the phase differences become too small to detect as the wavelengths of the sound become shorter. Information that comes from one side of the head will be more intense in the nearer ear and less intense in the further ear. The head is in the way of the further ear and absorbs high frequency sounds. The sonic shadow this produces is illustrated in *Fig. 1*. Cells in the **lateral superior olive** and the auditory cortex can detect these intensity differences.

These two mechanisms are complementary, providing a comprehensive system for the localization of sounds. Low frequency sounds produce phase differences but do not produce great differences in intensity at the two ears. Higher-frequency sounds produce poor phase differences but create a strong sonic shadow. The combination of both mechanisms allows for the detection of

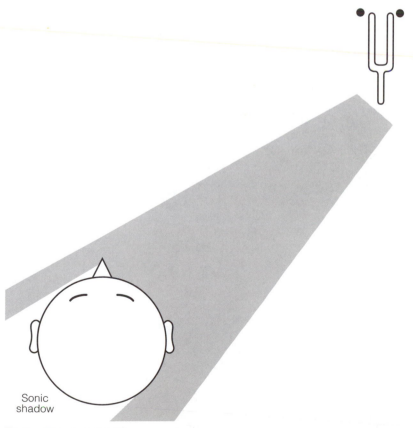

Sonic
shadow

Fig. 1.   The sonic shadow.

the location of sound across the whole frequency range and is known as the **duplex theory**.

Monaural cues

People who are deaf in one ear are not completely unable to localize sound. Monaural cues to the location of sound come from the pinnae of the outer ear. The pinnae dampen and filter the sound that enters the ear canal (known as **spectral filtering**). The modulations that occur will depend in part on the direction that the sound is coming from. Our ability to detect these differences gives us a localization cue. We need to appreciate here that our ability to recognize these changes alters as our head and ears grow from childhood into adulthood.

Differentiating 'in front' from 'behind'

Two sounds that are located an equal distance from your ears in the midline, where one is in front and one is behind, will be difficult to distinguish. There will be no difference of phase or latency, and the intensities of the sounds at the two ears will be identical. The knowledge of whether the sound is coming from in front or from behind relies on the pinnae of the outer ear. The differences in reflection of sound by the folds of the pinnae will differ for the two sound sources. Interestingly, sound location is best in the horizontal plane and much poorer for the near–far, in front–behind, and up–down directions.

I1 THE SOMATOSENSORY SYSTEM

Key Notes

The senses of touch	The skin contains several types of specialized receptor, and free nerve endings. Pacinian corpuscles are responsive to vibration. Merkel's disks and Ruffini endings are sensitive to pressure. Free nerve endings around hair roots respond to movement of the hairs by objects brushing the skin surface. Meissner's corpuscles are especially frequent in the fingertips and make fine discriminations. Information from these sensory systems passes to the brain through the dorsal-column medial-lemniscus pathway, ending up in the somatosensory cortex.
Pain	Pain receptors (nociceptors) are free nerve endings. They respond to tissue damage, which produces chemicals including potassium ions, bradykinin, histamine, and prostaglandins, which stimulate or sensitize nerve endings. Aspirin acts by preventing the synthesis of prostaglandins. Pain information is carried to the brain by fast A fibers and slow C fibers, and passes to the cingulate cortex, which is involved in emotion. Emotion and cognition can control the experience of pain. Melzack and Wall's gate-control theory holds that descending neurons carry information from the brain to synapses with A and C fibers, opening or closing a 'gate' for pain signals. Morphine acts on receptors in the periaqueductal gray matter in the center of the midbrain which naturally respond to endorphins.
Temperature	Temperature receptors seem to be free nerve endings. Sensations of warmth and cold are relative, depending on the adaptation of the skin. This makes temperature sensitivity difficult to study.
The vestibular system	The vestibular system in the inner ear provides information about our spatial orientation and movement through space. It comprises three semicircular canals, aligned in the three major axes of the body, and two otolith organs: the utricle and the saccule. These organs are stimulated when the head changes its orientation. Each semicircular canal has a swelling near its base; the ampulla, containing a sensory system that is stimulated when the head moves. Some information goes to the cerebellum, which concerns balance and spatial orientation. Other neurons pass to nuclei that control the eye muscles, producing automatic compensation for head movements (the vestibular–ocular reflex).
Related topics	Muscles and reflexes (J1) Homeostasis (L1)

The senses of touch

Vision (Section G) and audition (Section H) are **exteroceptive** senses, giving us information about the external environment. They are also *distance* senses, in that they respond to stimuli and events not in contact with our bodies. There are several skin senses, or senses of touch. These are also exteroceptive, but they are *contact* senses, giving us information about changes at the surface of the body: the skin.

Through the skin we are receptive to various types of stimulation. This is achieved through a number of types of specialized receptor, as well as free nerve endings. *Fig. 1* is a cross-section of the skin showing some of these receptors. *Table 1* summarizes these and their functions. Some of these receptor types occur in hairy skin, others in **glabrous** (smooth) skin, such as on the palms of the hands, and others in both. Glabrous skin contains more receptors, closer together. This reflects our use of the hands and fingers to actively manipulate the environment. Some of these receptor types are found in other organs of the body, giving sensitivity to the same kinds of stimulation as we receive in the skin.

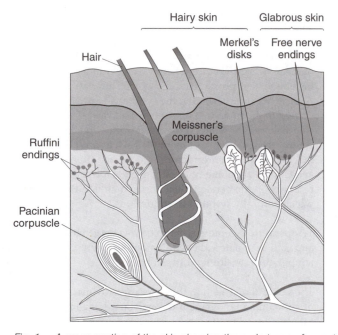

Fig. 1.   A cross-section of the skin showing the main types of receptor.

Table 1. The main sensory structures in the skin, and their relation to other skin structures

Receptor type	Type of skin	Layer	Function
Free nerve endings	hairy and glabrous	dermis; epidermis; around hair roots	temperature; pain
Pacinian corpuscles	hairy and glabrous	dermis; subdermis	high-frequency vibration
Ruffini endings	hairy	dermis	pressure or low-frequency vibration
Meissner's corpuscles	glabrous	dermis	low-frequency vibration
Merkel's disks	glabrous	epidermis	skin indentation

Stimulation of the skin by vibration or sudden movement of its surface causes the outer layers of **Pacinian corpuscles** to bend relative to the neural axons in their centers. This causes an action potential on both bending and unbending, making these receptors responsive to vibration but not to constant pressure. **Merkel's disks** and **Ruffini endings**, in contrast, respond to stimulation with sustained firing, making them sensitive to pressure or slow changes in pressure. Free nerve endings around hair roots respond to movement of the hairs by objects brushing the skin surface. **Meissner's corpuscles** are especially frequent in the fingertips. They produce an 'on' response to sharp edges, showing their role in fine tactile discriminations. These differences between types of response are sometimes described as showing *slow* (in the case of Merkel's disks and Ruffini endings) or *fast* (Pacinian corpuscles and Meissner's corpuscles) adaptation.

Information from these sensory systems passes to the brain mainly through the **dorsal-column medial-lemniscus pathway** (or more simply the **lemniscal system**). The sensory neurons have their cell bodies in the **dorsal root** of the spinal cord (see Topic D1), and the axons extend up the dorsal column to synapse with cells in the **medulla oblongata** (see Topic C1). Axons from these cells cross over to the other side and ascend to the **thalamus**. There they synapse with neurons whose axons pass to the **somatosensory cortex**, located in the postcentral gyrus. Penfield and Boldrey (1937) stimulated the cortex of humans undergoing brain surgery, and asked them to report what they felt. They showed that the regions of the body are represented in the somatosensory cortex in a systematic way, as shown in *Fig. 2*. More sensitive parts of the body occupy a larger area of the cortex. Later studies showed that different types of somatosensory information are represented in parallel strips along the length of the somatosensory cortex (Kaas et al., 1981).

Pain

The receptors for pain are free nerve endings distributed throughout the skin as well as in muscles, joints and internal organs. These are called **nociceptors**. They respond to stimulation that damages tissues, including mechanical injury (e.g. cutting, crushing), chemical damage (e.g. corrosive substances) and extremes of temperature (burning or freezing). Within the tissues, the nerve endings are stimulated by various chemicals produced by or released from damaged tissues. These include **potassium ions** from damaged cells, **bradykinin** from blood plasma leaking from damaged blood vessels, and **histamine** from mast cells, which form part of the mechanism of inflammation. Damaged cells also produce **prostaglandins**, which seem to sensitize nerve endings to other substances. The analgesic (pain-relieving) drug **aspirin** acts by preventing the synthesis of prostaglandins.

Pain information is carried to the brain by two types of neuron. Relatively thick, myelinated **A fibers** are responsible for the sharp initial pain in response to injury. Thin, unmyelinated **C fibers** carry information more slowly, and are responsible for dull, longer-lasting pain, of the type that often follows an initial acute pain. Both types of fiber convey information to the central nervous system through an **anterolateral system**. The cell bodies of the afferent neurons are, as with the touch senses, in the dorsal roots of the spinal cord. However, they synapse at the same level of the cord with neurons which cross to the other side immediately. Some of these fibers pass to the **thalamus**, while others synapse in the **reticular formation** (involved in arousal; see Topic K3) or elsewhere in the brain stem, from where other neurons pass the information to the thalamus.

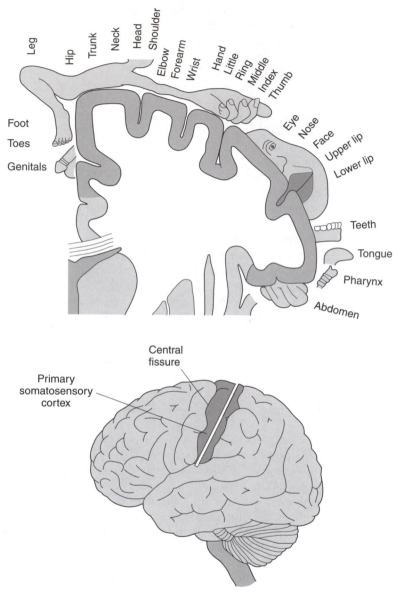

Fig. 2. The somatosensory cortex and the 'sensory homunculus', showing the relative areas of cortex concerned with each part of the body.

While pain information is clearly represented in the cortex (otherwise we would not be aware of it), it is not localized in the same way as tactile information. Pain information is known to be passed from the thalamus to the **cingulate cortex**, which is part of the **limbic system**. As we will see in Topic O2, this is involved in emotional experience and behavior. Pain is usually described as having affective (emotional) as well as sensory components.

Emotional and 'higher-level' cognitive processes can control the experience of pain. For example, if a person in acute pain is given an injection of pure water, believing it to be morphine, their pain will be dramatically relieved. This is an example of a **placebo effect**. Similarly, pain can be made worse by fear, for example if the sufferer believes it to result from a life-threatening illness rather than a trivial disorder. Melzack and Wall (1965) proposed a **gate-control theory** to explain such top-down effects on pain perception. The core of the theory is that stimulation of A fibers inhibits the firing of C fibers, thereby effectively closing a 'gate' for pain signals. This stimulation can come from cutaneous sensations, and is the basis for pain reduction when nearby areas of skin are strongly stimulated, for example by vigorous rubbing. The stimulation of the A fibers can also come from descending neurons from the brain, which is the route for the top-down influences of pain. One such pathway originates in the **periaqueductal gray (PAG)** region of the midbrain. Stimulation of the PAG reduces pain sensation. Analgesic drugs such as morphine act on receptors in the PAG. It is supposed that **endorphins**, which are produced by the body, reduce pain by acting on these receptors.

Temperature

The mechanisms through which we detect temperature are less well understood than other skin senses. The receptors seem to be **free nerve endings**. However, sensations of warmth and cold are relative, and depend on the adaptation of the skin to existing temperatures. The classic demonstration of this is to place one hand in hot water and the other in cold water. The sensations of heat and cold will gradually subside as the receptors adapt. If, after half a minute or so, both hands are then placed in the same bowl of water at room temperature, the water will feel cool to the hand that has adapted to hot water, and warm to the hand that has adapted to cold water.

This adaptation is one reason why temperature sensitivity is difficult to study. Another reason is because heating or cooling the skin changes its metabolic activity, and this can mask changes in neural firing, for example, due to changes in stimulation. Temperature information passes to the brain along the same **anterolateral system** as pain information.

The vestibular system

As well as sensory systems to tell us about the outside world, we have systems to tell us about our own position and motion within the environment. Together, these systems are called **proprioception**. Some proprioceptive information comes from receptors in muscles and joints (see Topic J1). Here we will look at the **vestibular system**, which is specialized for providing information about our spatial orientation and our movement through space.

The **vestibular organs** are located in the inner ear (see *Fig. 3*), alongside the cochlea (see Topic H1). They consist of three **semicircular canals**, which are aligned approximately in the three major axes of the body (see Topic C1), and

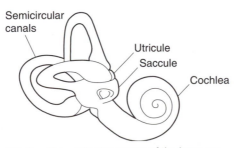

Fig. 3.   The vestibular organs of the inner ear.

two otolith organs, the **utricle** and the **saccule**. The word *otolith* literally means 'ear-stone'. Within each of these organs are crystals of calcium carbonate embedded in a gelatinous mass, the **otolith**. Inserted into this mass are **cilia** that are extensions of hair cells in the floor of the organ. As the head changes its orientation, the weight of the otolith pulls on the hair cells, stimulating them. In this way, the otolith organs provide information about changes in the orientation of the head.

Each semicircular canal has a swelling near its base: the **ampulla**. The whole structure is filled with liquid: **endolymph**. In each swelling is a gelatinous mass called the **cupula**, and in each cupula are cilia extending from receptive cells in the floor of the ampulla. If the head rotates in any direction, the semicircular canals move in the same direction. Inertia causes the endolymph to lag behind, pulling the cupula, which stretches the cilia and stimulates the hair cells. The relative stimulation of the three ampullas gives precise information about the direction of rotation. Normally, we are not aware of stimulation of these proprioceptive systems.

Information from the vestibular system passes to the CNS along the **vestibular branch** of the **auditory nerve**. Most of the axons from the hair cells pass to the **medulla**, but some go to the **cerebellum**, which, amongst other functions, is concerned with balance and spatial orientation. From the medulla, neurons pass to various brain sites, including other cranial nerve nuclei. These include the nuclei that control the eye muscles, so that changes of head position are automatically compensated by contrary eye movements, maintaining fixation on the same point. This is called the **vestibular–ocular reflex**.

Our most important proprioceptive sense is vision (see Section G). Many studies have shown that moving visual stimulation can overrule the other sources of proprioception. For example, when a person views a large-scale moving stimulus, after a short time the stimulus appears to stop moving, and instead the person feels they are moving in the opposite direction. The visual stimulus overpowers information from all the other sources that should indicate the person is stationary (Reason, Wagner and Dewhurst, 1981). An everyday example of this is the sense we sometimes have that the train we are sitting in is moving, when in fact it is an adjacent train that has started to move.

12 TASTE AND SMELL: THE CHEMICAL SENSES

Key Notes

Gustation	Gustation has four main sensory qualities: *bitter*, *sour*, *sweet* and *salt*. Receptors are located principally in papillae on the tongue, each containing several taste buds with several receptor cells. Papillae on different parts of the tongue carry taste buds sensitive to different tastes. Receptors are stimulated when chemical substances dissolve in the saliva coating the taste buds and come into contact with the gustatory receptors. Directly or indirectly, the substances open one of the ion channels in the receptor, depolarizing it.
Olfaction	We can distinguish thousands of odors, using about one thousand types of receptor. Receptors are in the olfactory epithelium, at the top of the nasal cavity, and are stimulated by airborne molecules that reach them through the nose or from the back of the mouth. Each receptor has cilia, which extend into the mucous layer of the epithelium, and an axon, which passes through the cribriform plate into the two olfactory bulbs. Scented substances dissolve in the mucous, where molecules attach to specific receptor molecules in the membrane of particular receptor cells, opening sodium channels and depolarizing the receptor.
Related topic	Dietary choice and psychological factors (M3)

Gustation

We prefer to use the word **gustation** for the sensory system located in the mouth, because, as we will see later in this Topic, what we mean by *taste* is largely based on olfactory stimulation. Both gustation and olfaction are extero-ceptive senses, giving us information about the outside world. They are *chemical senses* because the detection of stimuli and the triggering of the nerve impulse depend on chemical stimulation of the receptors. Gustation principally tells us about the properties of potential food which we put in our mouths. Gustation has four main sensory qualities: *bitter*, *sour*, *sweet* and *salt*. These qualities are important in dietary choice and dietary learning (see Topic M3). They tell us that some food may be potentially dangerous (bitter and sour), what food is good (sweet) and what contains minerals (salt). Evidence is becoming strong for a fifth quality, *umami*, the taste conferred by, for example, the monosodium glutamate (MSG) used in some Eastern cooking (for example in soy sauce).

Receptors for these sensory qualities are located throughout the mouth, but principally on the tongue. On the tongue, in particular, the sensory cells are located in **papillae**, each of which contains several **taste buds**. In turn, each taste bud contains several receptor cells. Papillae on different parts of the tongue carry taste buds sensitive to different tastes. The front of the tongue is maximally sensitive to sweet, the next region back, extending down the sides, to salt, the sides behind this to sour, and the central, back part to bitter.

Receptors are stimulated when chemical substances dissolve in the saliva coating the taste buds and coming into contact with the gustatory receptors. Salty substances contain metallic ions, usually sodium (Na^+), which enters a sodium channel and depolarizes the cell. Sour substances are usually acid, and contain hydrogen ions (H^+), which close potassium channels, producing depolarization. Bitter substances bind to a receptor on the receptor cell membrane, and this closes potassium channels, depolarizing the membrane. Sweet substances similarly bind to a receptor, but in this case they probably open calcium channels, depolarizing the membrane. Substances that give the umami sensation are certain amino acids. These probably bind to membrane proteins in specific types of receptor and open calcium (or nonspecific ion) channels, depolarizing the receptor membrane (Brand, 2000).

Gustatory information is conveyed along cranial nerves to the **nucleus of the solitary tract** in the **medulla**. From there, neurons travel to the **thalamus**, and from there to the primary sensory region of the **cerebral cortex**, the **amygdala** and the **hypothalamus**.

Olfaction

The other chemical sense is the sense of smell: **olfaction**. We can distinguish thousands of odors, and one basic question is: do we have a small number of receptor types, combinations of which produce this huge number of odors (compare color vision, Topic G1), or is there one receptor type for each odor? The answer seems to lie in between: we have about one thousand types of receptor.

These receptors are cells in the **olfactory epithelium**, two patches of tissue at the top of the nasal cavity (see *Fig. 1a*). This tissue is separated from the brain by a thin, perforated area of skull, the **cribriform plate**. The receptors are stimulated by airborne molecules, which reach the olfactory epithelium either through the nose or from the back of the mouth. As very little inhaled air normally reaches the top of the nasal cavity, in order to actively smell something we sniff, which carries more air to this region. Odors are also released by food in the mouth, and the molecules travel to the receptors from the back of the mouth. It is these molecules that provide most of what we describe as *flavor* or *taste*.

Each receptor cell has cilia, which extend into the mucous layer of the epithelium, and an axon, which passes through the cribriform plate into one of the two **olfactory bulbs** (see *Fig. 1b*). There they synapse with **mitral cells**, which pass information to the **temporal lobes**, ending in the **amygdala** and the **piriform cortex**. From there, neurons travel to other cortical areas and to the thalamus. The involvement of the amygdala, which is known to be involved in emotional control and memory (see Topic O2), explains why odors are such powerful evokers of emotion.

An odor is detected when a molecule of a substance reaches the olfactory mucosa. For a substance to have a scent it must dissolve in the mucous, where molecules attach to specific receptor molecules in the membranes of particular receptor cells. This opens sodium channels and depolarizes the receptors.

Many mammals have a second olfactory system, mediated by the **vomeronasal organ**, which is a group of receptors nearer the front of the nasal passage. This passes axons to **accessory olfactory bulbs** near to the olfactory bulbs. This system mediates the effects of **pheromones**, and is stimulated, not

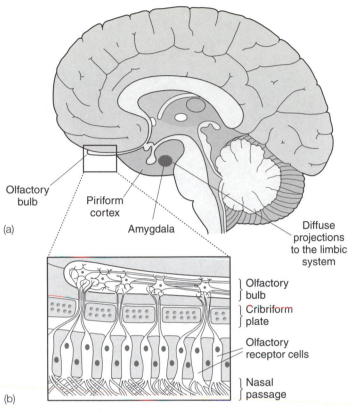

(a)

Olfactory bulb

Piriform cortex

Amygdala

Diffuse projections to the limbic system

} Olfactory bulb

} Cribriform plate

Olfactory receptor cells

} Nasal passage

(b)

Fig. 1 (a) The olfactory sensory system in relation to the brain. (b) The olfactory epithelium and its connections to the olfactory bulbs.

by airborne molecules, but by substances in urine and other secretions. In many species, these play an important role in social and reproductive behavior (see Topic N3). Russell, Switz and Thompson (1980) argued that the tendency of women living together to gradually synchronize their menstrual cycles is due to pheromones. However, there is doubt that such synchrony occurs in reality (e.g. Wilson, 1992). While there is evidence for the existence of a vomeronasal organ in human fetuses, most workers consider it not to be functional in humans (Takami, 2002).

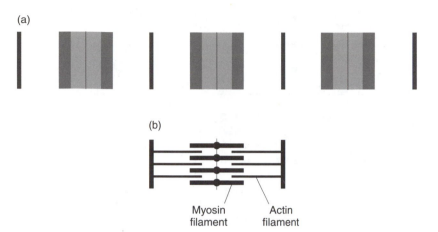

Fig. 1. Skeletal or striated muscle. (a) The general appearance of muscle. (b) The contractile system that gives its striped appearance.

at each end of a sarcomere being pulled into the central myosin filaments. This shortens the muscle, and is called *muscular contraction*.

Within each muscle are subsets of fibers, which are innervated by a single motor neuron. The motor neuron and the subset of fibers are together called a **motor unit**. Muscles that control fine movements, such as those controlling the fingers and eye movements, have small motor units (i.e. with few muscle fibers in each). Muscles that control gross movements, such as the biceps controlling flexion of the elbow, have larger motor units. Skeletal muscles do not contract in an all-or-none manner. Stronger contraction of a muscle results from the firing of more motor neurons, and hence the recruitment of more motor units.

Many muscles or muscle groups form opposed pairs with other muscles. Thus, for example, the *biceps*, which flexes the elbow, is opposed by the *triceps*, which extends it.

Neural control of skeletal muscle

As we just saw, muscles are controlled by **motor neurons**. When the motor neuron reaches the muscle it branches, innervating the fibers of its motor unit. Between the end of each branch of the neuron and the muscle fiber it innervates is a type of **synapse** (see Topic B3). **Acetylcholine** acts as the neurotransmitter, and attaches to receptors on the **motor endplate** of the muscle fiber. This starts an action potential, which in turn causes muscle contraction through the myosin–actin mechanism described earlier. Each motor neuron leaves the ventral horn of the spinal cord, although, as we shall see below and in Topic J2, there are two main sources of stimulation of motor neurons.

Each muscle contains **intrafusal muscle fibers**, also called **muscle spindles**, which are stretch receptors. When a muscle is stretched by contraction of an opposing muscle, the afferent nerve fiber serving each spindle passes information to the spinal cord. Stretch receptors in the tendons which attach muscles to bones, the **Golgi tendon organs**, also serve as stretch detectors, passing afferent signals to the spinal cord. However, whereas intrafusal fibers respond to passive stretch of the muscle, Golgi tendon organs respond proportionally to the tension in the muscle. Both types of receptor are involved in the reflexes we examine next.

J1 MUSCLES AND REFLEXES

Key Notes

Striated muscle	Striated (or skeletal) muscle is composed of parallel muscle fibers. Within each fiber are numerous myofibrils, which run the full length of the fiber. The stripes in striated muscle are sarcomeres. A sarcomere is composed of filaments of actin and myosin. Muscle contraction results from the actin filaments being pulled into the myosin filaments. A single motor neuron innervates a group of fibers, forming a motor unit.
Neural control of skeletal muscle	The neurotransmitter across the synapse between the motor neuron and the muscle fiber is acetylcholine. Each muscle contains stretch receptors called intrafusal muscle fibers (muscle spindles). These and Golgi tendon organs serve as stretch detectors, and are involved in reflexes.
Reflexes	Stretching a muscle stimulates stretch receptors. Sensory neurons enter the dorsal root of the spinal cord and synapse with motor neurons. This monosynaptic stretch reflex acts as a continuous feedback loop matching muscular contraction to muscle load. Polysynaptic reflexes involve more than one synapse. Axons from pain receptors synapse in the spinal cord with interneurons, which connect with several motor neurons, producing coordinated withdrawal. Some interneurons pass to the other side of the body, producing the crossed extensor reflex. Reciprocal innervation allows opposed groups of muscles to organize their contraction and relaxation.
Smooth muscle and cardiac muscle	The actions of many internal organs are carried out by smooth muscle. It is innervated by both branches of the autonomic nervous system, which have opposing actions. Cardiac, or heart, muscle is a specialized tissue that causes the heart to contract.
Related topic	Control of movement by the brain (J2)

Striated muscle   All movement is the result of the contraction of muscles. The muscles that we feel under our skins, as well as many internal muscles that we cannot feel, such as the *extraocular muscles* that move the eyeballs, are made of contractile tissue called **skeletal** or **striated muscle**. A muscle is a bundle of numerous **muscle fibers**. Each muscle fiber is a single long cell, and the cells in skeletal muscle are arranged parallel to each other, but are separate. Within each fiber are numerous **myofibrils**, which run the full length of the fiber. The name 'striated muscle' comes from the striped appearance that skeletal muscle has under the microscope (see *Fig. 1*). The stripes consist of a number of bands and lines in each myofibril. Each set of bands and lines is a functional unit called a **sarcomere**. A sarcomere is composed of intersecting filaments made of proteins called **actin** and **myosin**. When a muscle is stimulated, a contraction occurs as a result of the actin filaments

Reflexes

While much of our muscular activity is controlled by the brain (see Topic J2), many very important muscle movements are controlled by **spinal reflexes**.

Reflexes are actions that take place automatically in response to some stimulus, and permit immediate actions without the intervention of the brain. The simplest reflex is the **monosynaptic stretch reflex**. The best-known example of this is the *patellar reflex*; the knee jerk used by physicians to test the activity of spinal nerves. In this example, striking the tendon below the knee cap (patella) stretches the quadriceps, the extensor muscle in the front of the thigh. Stretching the muscle stimulates the muscle spindles. Sensory neurons from the muscle spindles enter the dorsal root of the spinal cord, and synapse with motor neurons in the ventral root, which causes the muscle to contract. The normal function of this reflex is to cause a quick adjustment of muscle contraction when the load on the muscle is increased. It acts as a continuous feedback loop matching muscular contraction to the load on the muscle.

Polysynaptic reflexes are spinal reflexes that involve more than one synapse. An example is the **withdrawal** or **flexion reflex**, when a limb is rapidly removed from a painful stimulus. In this reflex, pain receptors send signals along axons to the dorsal root of the spinal cord (see Topic I1). These axons synapse with a number of short **interneurons**. These interneurons connect with several motor neurons, producing a coordinated withdrawal of the affected part of the body. However, some of the interneurons connect with motor neurons on the other side of the body, where they produce limb extension: the **crossed extensor reflex**. The importance of this is that if, for example, I stand on a pin, the resulting reflex withdrawal of that foot will be compensated by greater support in the other leg.

Another type of reflex results from **reciprocal innervation**. This allows opposed groups of muscles to organize their contraction and relaxation, permitting coordinated actions such as walking. The simplest form of reciprocal innervation is when the flexor and extensor muscles around a joint work together. In walking, for example, the knee joint bends because the quadriceps muscle relaxes as the hamstring muscle flexes. Notice that this is not a sudden on–off action; contraction and relaxation are gradual and coordinated. The use of both legs in walking involves this reciprocal innervation, and also the alternation of the actions of the two legs by means of the crossed extensor reflex. All of this occurs without the necessity for conscious intervention.

The main principles of reflexes were demonstrated around the end of the 19th and beginning of the 20th centuries by Charles Sherrington. He showed how, in many organisms, much behavior is composed of chains of reflexes. With increasing encephalization (see Topic A1) these are increasingly controlled by brain mechanisms (see Topic J2).

Smooth muscle and cardiac muscle

The actions of many internal organs are carried out by **smooth muscle**. Smooth muscle underlies many of the physiological changes we will describe in later sections of this book. Amongst other places, smooth muscle is found in:

- the walls of blood vessels, where it is responsible for their constriction and (by relaxing) dilatation. This permits fine control of blood flow to different tissues and organs;
- the walls of the intestines, where it is responsible for the contractions, called *peristalsis*, which move food through the gastrointestinal tract;
- the sphincters that control filling and emptying of the gastrointestinal tract and the genitourinary tract;

- the iris and the lens muscles within the eye;
- the hair follicles of the skin, where it produces *piloerection* (raising of the hairs) in response to cold.

Generally, smooth muscle reacts slowly to stimulation, and produces mostly longer-lasting changes. Smooth muscle is innervated by both branches of the autonomic nervous system (see Topic D2), which have opposing actions. As demands on the body vary, the parasympathetic and sympathetic nervous systems vary the contraction of the smooth muscle, matching the state of the organ to those demands.

Cardiac, or **heart**, **muscle** is a specialized tissue that causes the heart to contract (beat), pumping blood through the blood vessels. It, too, is controlled by opposing actions of both branches of the autonomic nervous system.

J2 CONTROL OF MOVEMENT BY THE BRAIN

Key Notes

The motor cortex

Individual movements are controlled by the primary motor cortex in the precentral gyrus. The main inputs to this cortex come from the premotor cortex and the supplementary motor area. These produce movement programs. The planning of more complex behavior takes place in the prefrontal cortex. Lesions in the association areas involved in movement produce a variety of apraxias: disorders of skilled movement not due to sensory loss or paralysis.

Motor pathways

The pyramidal system has two main components: the corticospinal tract, consisting mainly of axons from cortical neurons which synapse with motor neurons. The corticobulbar tract passes to various cranial nerves. Most of the neurons originate in large pyramidal cells in the primary motor cortex. Lesions of the pyramidal system cause voluntary paralysis. The other motor pathways have collectively been called the extrapyramidal system. Lesions in these tracts produce difficulties involving excessive movements or limited movement. The extrapyramidal system includes the vestibulospinal tracts, the rubrospinal tract and the reticulospinal tracts. The motor system is an integrated one, including both voluntary and involuntary components.

The basal ganglia

The basal ganglia receive inputs from the motor and the somatosensory cortices, and project to the secondary and primary motor cortices. Destruction of parts of the basal ganglia causes serious motor disorders. Parkinson's disease, resulting from degeneration of neurons in the substantia nigra, is characterized by a resting tremor. Huntington's chorea produces uncontrollable jerky movements. It results from degeneration of the striatum.

The cerebellum

The cerebellum receives inputs from the motor cortex, the vestibular system, the special senses, and the somatosensory system. Its outputs go to the extrapyramidal system. Damage to the cerebellum causes deficits in the ability to produce smooth movement, including an 'intention tremor'.

The brain stem

Many automatic processes controlled by the extrapyramidal system, and species-typical sequences of behavior, are programmed in regions of the brain stem. These programs are responsive to sensory inputs and conscious modification.

Related topic

Muscles and reflexes (J1)   The neural basis of language (Q2)

The motor cortex   Individual movements are controlled by the **primary motor cortex**. This occupies the **precentral gyrus**, parallel to the somatosensory cortex in the postcentral gyrus (see Topic I1). Penfield and Rasmussen (1950) showed how each part of the body is controlled by a separate area of the motor cortex, and parts capable of the most precise movements (especially the fingers and lips) have larger areas of cortex devoted to them. This can be represented by a *motor homunculus* similar to the sensory homunculus shown in *Fig. 1* of Topic I1. There are direct connections between neurons in the somatosensory cortex (see Topic I1) and in the motor cortex that serve the same parts of the body. Evarts (1974), recording from the primary motor cortex of monkeys, showed that tactile stimulation of the hand produces very rapid responses in corresponding motor neurons, presumably by way of these connections.

The main inputs to the primary motor cortex come from the **premotor cortex** and the **supplementary motor area** located in the gyrus anterior to the primary motor cortex. Together, these are sometimes called the **secondary motor cortex**. Their function seems to be to produce movement programs: coordinated sets of movements. Through numerous connections with the primary motor cortex, these motor programs control individual movements, and produce coordinated behavior. Studies using PET scans in humans have shown that the primary and secondary motor cortices, as well as the primary somatosensory cortex, are all active both during performance of well-learned response sequences and while learning new sequences (Jenkins et al., 1994). In addition, during the learning of new sequences, but not during well-practiced ones, part of the **prefrontal cortex** is active. Together with lesion studies, this suggests that the planning of more complex behavior, or the construction of motor programs, takes place in the prefrontal cortex. (In Topic Q2 we will look at this in more detail in relation to speech.)

The prefrontal cortex and the secondary motor cortex both receive inputs from areas of association cortex involved with sensory and spatial information processing. Lesions in the primary motor cortex produce paralysis of particular muscles or muscle groups. However, lesions in the other areas involved in movement produce a variety of **apraxias**: disorders of skilled movement not due to sensory loss or paralysis. For example, patients may be unable to walk smoothly or follow instructions to move their legs (**limb apraxia**). In **construction apraxia** a person may be unable to make the coordinated movements necessary to copy simple designs or complete a jigsaw puzzle.

Motor pathways

We saw in Topic J1 that the contraction of muscles is controlled by motor neurons in the ventral roots of the spinal cord. The pathways by which the brain controls these motor neurons are complex, and what follows is a simplification.

One major pathway is the **pyramidal system**. In one part of this, the **corticospinal tract**, axons from cortical neurons pass through the brain stem and down the spinal cord before synapsing with motor neurons. Most of these axons *decussate* (cross to the other side) in the brain stem, the remainder do so in the part of the spinal cord where their target motor neurons are located. The other part of the pyramidal system, the **corticobulbar tract**, passes to the medulla where it reaches the nuclei of various cranial nerves (see Topic C2).

Most of the neurons originate in large pyramidal cells in the primary motor cortex, although many originate in the primary somatosensory cortex, and some in association areas such as the premotor cortex. Some axons in the pyramidal tract end in various sensory nuclei in the brain stem and spinal cord, and in the reticular formation (see Topic C2). Historically, the pyramidal system was characterized as the pathway by which voluntary action is controlled; lesions confined to the pyramidal system cause paralysis of voluntary movement, while many reflexes remain intact. However, the separation of voluntary and automatic control is not actually this simple. Axons from the primary motor cortex also pass to the **basal ganglia** and to the **cerebellum** (see below), historically considered to be involved with non-voluntary movement, both of which then feed back to the primary and secondary motor cortex. The sensory connections indicate the reliance of voluntary movement on immediate sensory feedback.

The other motor pathways have collectively been called the **extrapyramidal system**, and characterized as involved in automatic (non-voluntary) movement. Lesions in these tracts do not cause the loss of particular movements. Rather, patients show a variety of difficulties involving either excessive movements that interfere with voluntary actions, or slowed or limited movement that interferes with both voluntary and involuntary movements. A number of tracts carry this sort of motor information, including the **vestibulospinal tracts**, which originate in the brain stem, and influence the tone of limb muscles. They receive inputs from the vestibular system, and help to control posture and maintain eye position. The **rubrospinal tract** has its origin in the **red nucleus** high in the brain stem. This receives inputs from the primary motor cortex and from the cerebellum (see below). The rubrospinal tract connects to motor neurons of the major limb muscles, so is involved in locomotion. The **reticulospinal tracts** originate in the reticular formation of the brain stem. Their functions are not well understood, but they promote certain reflexes, and help control automated actions such as walking and maintaining posture.

As we have said, we cannot fully separate voluntary and involuntary components of the motor system. The whole system is one in which automatic and voluntary pathways are integrated with each other and with sensory mechanisms.

The basal ganglia   The **basal ganglia** are a group of interconnected subcortical nuclei. They include the **caudate nucleus** and the **putamen** (together called the **striatum**), the **globus pallidus**, the **substantia nigra** and the **subthalamic nucleus**. The striatum receives inputs from many parts of the cerebral cortex, particularly the primary motor cortex and the primary somatosensory cortex. In turn, the striatum projects to the globus pallidus, and has two-way connections with the substantia nigra. The globus pallidus has two-way connections with the subthalamic nucleus, and projects to the secondary and primary motor cortices by way of the thalamus.

The functions of the basal ganglia are not fully worked out. In general, their effect is inhibitory on the thalamus and hence on the cortex. Destruction of parts of the basal ganglia causes characteristic, serious motor disorders. **Parkinson's disease** is a condition characterized by movement disorders including a tremor when not trying to move, slowness, and difficulty starting movements. The cause of Parkinson's disease is degeneration of dopaminergic neurons in the **substantia nigra**, which prevents information passing back to the striatum. Parkinson's disease is treated by large doses of **L-dopa** ('Levodopa'), a precursor of dopamine. This passes into the brain and into the remaining neurons in this circuit, enabling them to produce more dopamine to affect the striatum.

Huntington's chorea is a hereditary disease producing uncontrollable jerky movements of the face and limbs. It results from degeneration of the striatum, particularly of GABAergic neurons.

It is becoming clear that the basal ganglia are also involved in emotion. For example, Schneider et al. (2003) have shown that electrical stimulation of the subthalamic nucleus, used as a treatment for long-term sufferers from Parkinson's disease, increases their sense of well-being and improves their emotional memory.

The cerebellum   The **cerebellum** is a relatively large structure located in the hindbrain (see Topic C1). It contains more neurons than the cerebral cortex, and structurally has several **deep cerebellar nuclei** surrounded by a cortex. The cerebellum receives inputs from the motor cortex, from the vestibular system (see Topic I1), from the special senses and from the somatosensory system (see Topic I1). Its outputs go to motor centers in the extrapyramidal system. Damage to the cerebellum causes serious deficits in the ability to produce smooth movement, including an 'intention tremor' which contrasts with the resting tremor of Parkinson's disease. Other symptoms can include muscle weakness, lack of coordination, slurring of speech, and a staggering gait (ataxia). These all suggest that the cerebellum plays a key role in programming sequential behaviors, especially those requiring integration of external stimuli and timing. The cerebellum is involved in motor learning (see Ohyama et al., 2003)

More recent studies have suggested that the cerebellum is important in emotional and cognitive processes. For example, Leroi et al. (2002) reported very high rates of psychiatric illness in patients with cerebellar disease.

The brain stem   A large class of behaviors that fall between the intentional, voluntary movements controlled by the cerebral cortex and the reflexes controlled by the spine are programmed by various centers in the brain stem. These behaviors include the automatic processes controlled by the extrapyramidal system, such as postural changes, eye movements, breathing, walking and the like. They also include more complex, species-typical sequences of behavior that serve basic biological needs. Such preprogrammed sequences can be seen in domestic animals — for example, the stereotypical posture and movements of a hunting cat. We will see examples of these in later sections (e.g. drinking, Topic L2; feeding, Topic M2; mating, Topic N1).

For some of these behaviors in other species, control centers have been identified in various parts of the brain stem. For example, a region of the **periaqueductal gray** (PAG) produces sexual receptive behavior in female rodents. Note that, although these species-typical behaviors are preprogrammed, they are not necessarily inevitable. They are responsive both to sensory inputs and to conscious modification.

K1 THE NATURE OF SLEEP AND DREAMS

Key Notes

Physiological activity during sleep	During sleep, activity in the electroencephalogram (EEG) slows to slow-wave (SW) sleep. The EEG shows several slowing and speeding cycles of about 90 minutes. EEG slowing is accompanied by muscular relaxation, slowing of heart rate and increased secretion of growth hormone.
Rapid eye movement sleep	During each sleep cycle a period of rapid eye movement (REM) sleep occurs, accompanied by an apparently alert EEG, relaxation of muscles, and faster breathing and heart rate. PGO waves and hippocampal theta activity can be recorded at the onset of REM sleep in animals.
Sleep across species	Sleep becomes more complex further up the phylogenetic scale. Birds and mammals show REM and SW sleep. Different species spend different amounts of time asleep, and different proportions in REM sleep. In mammals, REM sleep correlates with total sleep, and REM and SW sleep are negatively related to body weight. REM sleep is longer in altricial mammals, and species with secure sleep. Sleep in most mammals is polyphasic but primates have a monophasic sleep pattern.
Development of sleep	At birth, REM and non-REM sleep each occupy about half of total sleep. Neonates sleep 15–17 hours each day. SW sleep first appears at 4–5 months. Sleep decreases throughout life, and the proportion of non-REM sleep increases. In later life stages 3 and 4 sleep decline.
Dreaming	Dreams during REM are more vivid than those in non-REM, and often include elements from waking events. Episodic memory is probably inaccessible during both REM and non-REM sleep. People who claim that they never dream have just as many REM sleep periods as others, and are about as likely to report dreams when awoken during these periods.
Related topics	Mechanisms of sleep and arousal (K3) Disruptions of sleep and rhythms (K5)

Physiological activity during sleep

Until the mid 20th century sleep was generally considered to be a state of *minimal arousal*. This view changed after the introduction of the **electroencephalograph** (EEG; see Topic A2). Aserinsky and Kleitman (1953) showed that the EEG of a typical adult night's sleep shows a characteristic sequence (see *Fig. 1*). When we are awake and mentally active the EEG shows relatively fast and irregular **beta activity** (around 13–30 Hz), also called **desynchronized EEG**. When we close our eyes and relax this is replaced by the **alpha rhythm**, or **synchronized EEG** (around 8–12 Hz). **Stage 1 sleep** is characterized by slower

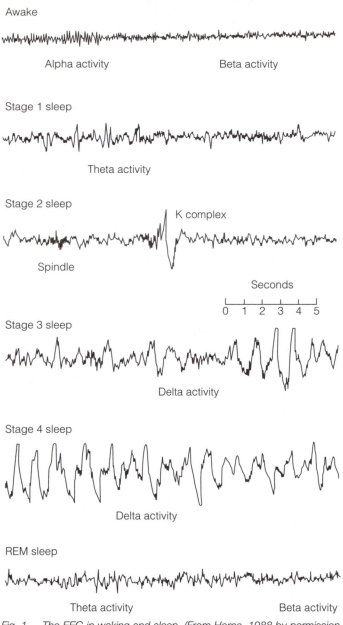

Fig. 1.   The EEG in waking and sleep. (From Horne, 1988 by permission of Oxford University Press.)

theta waves (3.5–7.5 Hz), and **stage 2 sleep** by irregular, mostly slow activity, interspersed with faster bursts (**sleep spindles**) and spikes (**K complexes**). **Stage 3 sleep** shows high amplitude **delta activity** (less than 3.5 Hz), and this is more marked in **stage 4 sleep**.

We go through a number of cycles of these stages during the night, with a period of about 90 minutes (see *Fig. 2*). The pattern of stages is called the **sleep architecture**. As the night goes on less time is spent in stage 4, and more in stages 2 and 3. The initial sequence from stage 1 through to stage 4 might take

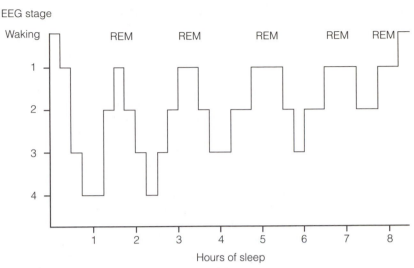

EEG stage

Fig. 2.   Typical pattern of sleep through one night.

only half an hour. Stage 1 is a transition between sleep and waking, and a person woken from stage 2 might claim not to have been asleep. We call stages 1–4 **non-REM sleep**, and stages 3 and 4 together are **slow-wave (SW) sleep**. Muscular relaxation and progressive slowing of heart rate accompany the EEG changes during passage to SW sleep. During the first few hours of sleep there is increased secretion of **growth hormone**.

Rapid eye movement sleep

After the first cycle to stage 4, rapid movements of the eyes almost always accompany occurrences of stage 1 sleep (Aserinsky and Kleitman, 1953). This stage, called **rapid eye movement (REM) sleep**, is associated with the deep relaxation of the muscles of the trunk found in SW sleep (Berger and Oswald, 1962), although there is likely to be twitching of the muscles of the limbs and face. However, there is faster breathing and heart rate. REM sleep is sometimes called **paradoxical sleep** because, while the EEG is similar to the waking EEG, it is accompanied by the muscle relaxation of SW sleep. It is also usually harder to awaken people from REM sleep than from SW sleep, despite the apparently alert EEG. In addition, REM sleep shows increased sympathetic nervous system activity (including increased and irregular heart rate) and increased genital blood flow (producing penile erection in males). Homeostatic control of body temperature (see Topic L1) is poor during REM sleep (Heller and Glotzbach, 1985).

At the onset of REM sleep in laboratory animals electrodes placed into several brain structures – the pons, the lateral geniculate nucleus and the occipital cortex – record characteristic bursts of spike discharges known as **PGO waves**. (These are assumed to exist in humans, although they cannot be observed in surface EEGs.) PGO waves are the first sign of REM sleep, and are followed by desynchronization of the EEG, the cessation of muscular activity and rapid eye movements. One other notable characteristic of REM sleep is **hippocampal theta**: rhythmic activity of 4–7 Hz that can be recorded from the hippocampus. The hippocampus is a key structure in certain types of memory (see Topics Q1 and Q2), and this activity might be related to the effects of sleep on memory consolidation (see Topic K2).

Sleep across species

Broadly, sleep increases in complexity as we go further up the phylogenetic scale. Insects, mollusks, crustaceans, amphibians and most fish show periods of relative inactivity, but do not have slowing of the EEG. Most reptiles have SW sleep but not REM sleep. Birds and mammals show both types of sleep, but different species spend different amounts of the 24 hours asleep, and different proportions of their total sleep in REM sleep. Some birds and some marine mammals show SW sleep in one hemisphere and a waking EEG in the other. Marine mammals have to surface to breathe, which they cannot do if fully asleep. Some species (e.g. seals) solve this by sleeping in short spells while they hold their breath (Lyamin, 1993), while others (e.g. sea lions and dolphins) breathe whilst asleep by keeping one hemisphere alert while the other is in SW sleep (Mukhamentov, Supin and Lyamin, 1988). In birds, it seems to help them avoid becoming the prey of another animal. Birds on the outside of a group are more likely to show unilateral sleep, and they keep the eye that faces away from the group open (Rattenborg et al., 1999).

REM sleep parameters are not related to characteristics such as body temperature, life span, brain size, or brain/body size ratio. However, within mammals, four major sleep time determinants have been established:

- REM sleep time is correlated with total sleep time;
- both REM and SW sleep duration are negatively related to body weight;
- REM sleep time is higher in **altricial** mammals (those, such as primates, born relatively immature), especially soon after birth;
- REM sleep is longer in animals with more secure sleep (either predators or those with safer resting sites).

These relations hold for all mammals studied except for the egg-laying *echidna*, whose phylogenetic line diverged from the other mammals some 130 million years ago (see Siegel, 1995).

Most mammal species exhibit **polyphasic** sleep; that is, they sleep a number of times (up to 12) in each 24-hour cycle (Campbell and Tobler, 1984). The remaining species, including humans and other primates, sleep only once each day: a **monophasic** sleep pattern. Human infants are polyphasic, and the monophasic pattern emerges during the first year. However, it may take years for daytime sleeping to disappear completely, and many adults continue to experience polyphasic tiredness, such as feeling sleepy in early afternoon.

Development of sleep

In humans, and other mammals, the total time spent sleeping and the proportion of time spent in REM sleep are both greatest before birth, and decrease with age (Roffwarg, Muzio and Dement, 1966; see *Fig. 3*). Studies of premature infants have shown that at between 24 and 26 weeks gestation (14–16 weeks premature) the EEG during sleep is flat, showing only sporadic activity. Then, periods of non-REM sleep increase, until REM and non-REM sleep occupy about 50% each at full term. SW sleep starts to appear at 4–5 months of age, and increases thereafter. The neonate spends some 15–17 hours asleep each day. This decreases to an average of 8 hours in youth, then down to 6–7 hours in the 50s and 60s.

The proportion of REM sleep also gradually declines from 50% at birth, through about 25% in young adulthood, and around 15% at age 50–60 (and in some individuals may eventually disappear altogether). The proportion of SW sleep during non-REM declines, and it may disappear altogether. While elderly

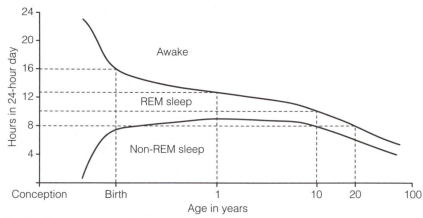

Fig. 3.   The life-span development of sleep in humans.

people may sleep less, they tend to awaken more often, tend to revert to polyphasic sleep, and complain more of insomnia (see Topic K5). When adults are put in situations isolated from environmental stimuli (see Topic K4) the tendency to sleep more frequently and to adopt a polyphasic sleep pattern returns.

Dreaming   Dement and Kleitman (1957) found that people woken during REM periods almost always reported vivid dreams. Waking in non-REM periods resulted either in no dream reports or in vague, easily forgotten reports. The characteristic rapid eye movements were thought to show the viewing of visual images in dreams. Subsequent research has shown more dream activity in non-REM sleep (Cartwright, 1979). Further, the eye movements in REM sleep are different from those in waking. Dreams during REM periods may be more vivid, show more bizarre elements and associations, and have a greater emotional content. They often also include elements from memories of waking events. However, they rarely replay the waking episodes from which the elements come, although they show a story-like (if bizarre) sequence. It is thought that **episodic memory** (memory for everyday events, as opposed to concepts and ideas) is inaccessible during both REM and non-REM sleep. External stimuli can become incorporated into dreams. For example, water sprayed on the dreamer often results in dream experiences of getting wet. People who claim that they never dream have just as many REM sleep periods as others, and are about as likely to report dreams when awoken during these periods. We seem only to remember a dream when we awaken during it, or very soon after.

K2 THE FUNCTIONS OF SLEEP

Key Notes

Total sleep deprivation	Prolonged sleep deprivation is accompanied by very little physiological change. Limited effects on vigilance tasks, and occasional paranoid thoughts, might be attributable to microsleeps. Rats die if deprived of sleep for about 4 weeks. They show a variety of pathological changes, suggesting that sleep serves a number of functions, including thermoregulation, energy conservation and immune system maintenance. REM sleep deprivation has no persistent effects different from those of general loss of sleep.
Recuperative and circadian theories	Recuperative theories view sleep as a period during which recovery or repair takes place. There is evidence for a general restorative function for SW sleep. Circadian theories suggest sleep evolved as a way of fitting organisms to the light–dark cycle. This fits better with some of the species differences in sleep pattern. Sleep periods relate to the need to obtain nourishment.
Adaptive theories	Developmental theories emphasize the predominance of REM sleep early in development. REM sleep might play a key role in brain development. Learning theories propose that REM sleep promotes the formation of long-term implicit or procedural memory. Post-training sleep deprivation impairs performance; REM sleep increases on post-training nights; neural activity during training is reinstated during sleep; brain areas that are more active during training task are also more active in subsequent REM sleep. Sleep influences the consolidation of implicit or procedural memory, but not explicit or declarative memory. Dreams might be by-products of sensory activation during sleep, or they might be the conscious representations of the reprocessing of memory.
Related topics	The nature of sleep and dreams (K1) Disruptions of sleep and rhythms (K5)

Total sleep deprivation

People often think it would be useful not to bother sleeping; that sleeping is a waste of time. But if we try to do without any sleep, the desire to sleep increases markedly and it is difficult to stay awake after the first 48 hours. This suggests that sleep is essential, and serves some important function or functions. In 1965 a 17-year-old student, Randy Gardner, stayed awake for over 264 hours. Later, the Guinness Book of Records recorded that Maureen Weston had stayed awake for 449 hours. Curiously, although sleep seems to be essential, prolonged sleep deprivation is accompanied by very little physiological change, and little interference with cognitive processes. Dement (1978), who observed Randy Gardner, reported that his behavior appeared normal, although Coren (1995) has reported that he actually had trouble focusing his eyes on day 2, hallucinations on day 4, and slurred speech and a short attention span by the last day. In general, performance of tasks involving reasoning, spatial relations and comprehension are usually found to be unaffected, especially when conducted against the clock. What does deteriorate is performance of simple mood and cognitive tests, and tasks involving vigilance or prolonged attention (Horne, 1978).

After about 60 hours of sleep deprivation, hallucinations and paranoid thoughts sometimes occur (Morris et al., 1960). These are not related to **psychosis**, since schizophrenic patients show normal sleep patterns, and their symptoms are not affected by sleep deprivation (see Topic R1). Many of these effects might be due to an increasing tendency to have **microsleeps** during sleep deprivation: that is, we have an increasing tendency to very brief periods of REM sleep whilst trying to stay awake. Following sleep deprivation, we recover only about 20–25% of the lost sleep on the following two or three nights, after which the sleep period returns to normal.

In 1983, Rechtschaffen and others developed a method of extending sleep deprivation in rats beyond that possible voluntarily in humans (see Rechtschaffen and Bergman, 1995). A **yoked pair** of rats is placed in cages on a turntable (see *Fig. 1*). EEGs are recorded, and when the EEG of the experimental rat shows that it is falling asleep, the turntable is rotated, tipping both animals into a water bath. Although the yoked control rat lands in the water equally often, this is not caused by its sleep pattern. So, while the experimental rat might get only 15% of its normal sleep, the control rat could get 70%. Typically, the sleep-deprived rats die after about 4 weeks, while the control rats show little or no ill effects. The deprived rats have a variety of pathological changes, but what they show in common is increased peripheral energy expenditure (with increased food intake, weight loss, and endocrine and enzyme changes associated with increased metabolic activity), coupled usually with decreased body temperature. They also have depressed immune function, making them susceptible to infection. These changes suggest that the functions of sleep include maintaining thermoregulation, energy conservation and immune system function.

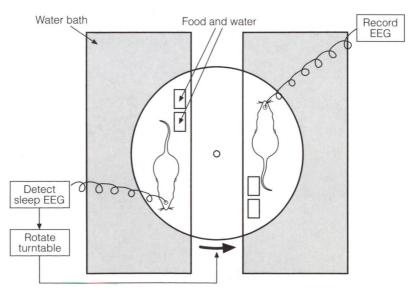

Fig. 1.   Diagram of the sleep carousel apparatus for investigating sleep deprivation in rats.

The effects of **selective deprivation of REM sleep** reveal a powerful drive to enter REM sleep (see *Fig. 2a*). If we wake people on successive nights as soon as their EEG shows that they are starting REM sleep, they enter REM sleep with increasing frequency, up to 50 times or more in a night (Webb and Agnew, 1967). When, at the end of the deprivation period, we let them sleep uninterrupted they spend as much as twice the usual amount of time in REM sleep, although the overall amount of sleep is hardly increased (Brunner et al., 1990; see *Fig. 2b*). While this suggests that REM sleep is important, there seem to be no persistent psychological effects of REM sleep deprivation different from those of general loss of sleep. Indeed, the case has been reported of a soldier who suffered a localized wound to the pons. He was found to have almost no REM sleep, yet showed no adverse effects.

(a)

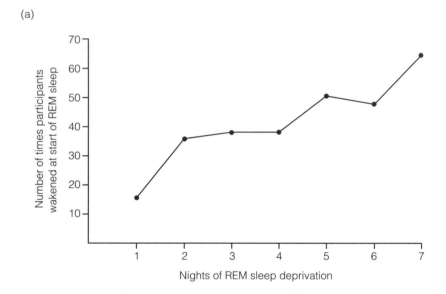

(b)

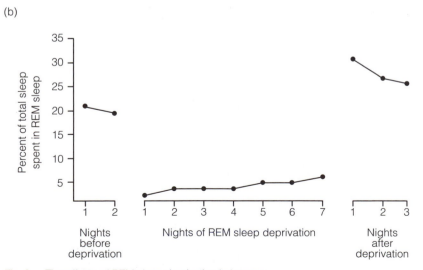

Fig. 2.   The effects of REM sleep deprivation in humans.

Recuperative and circadian theories

Many theories have been proposed to explain the functions of sleep. **Recuperative theories** view sleep as a period during which recovery or repair takes place. Recent versions have proposed processes to do with immune function or thermoregulation. One problem with such a general approach is that REM sleep is maladaptive in the homeostatic sense, since temperature control is greatly reduced.

There is evidence for a general restorative function for SW sleep. We know that we sleep well after exercise, and some studies have indicated that vigorous exercise selectively increases SW sleep. However, other studies have shown no such effect. Maybe only exercise that raises the temperature of the brain increases SW sleep (Horne, 1988). We know that local heating of the head without exercise increases subsequent SW sleep. Experience tells us that other circumstances that raise body temperature, such as fever and hot weather, induce sleepiness. The areas of the basal forebrain that control sleep are also involved in temperature regulation (see Topic K3).

Circadian theories suggest that sleep evolved as a way of fitting organisms to the light–dark cycle (see Topic K4). This would conserve energy for times of the day when animals need to be active to seek food, mates and so on, and allow them to hide at other times. Circadian views fit with some of the species differences in sleep pattern better than recuperative views (see Topic K1). While there is no relationship between daily sleep time and body size or activity level, sleep periods do relate to the need to obtain nourishment: grazing animals tend to sleep for short periods, and hunters for long periods. Humans fall in between.

Adaptive theories

Developmental theories of REM sleep emphasize its predominance early in development (see Topic K1). For example, they suggest it plays a key role in brain development (Roffwarg et al., 1966). This could be by promoting synaptic connections, or by controlling the neural activity involved in activity-dependent development. A related view is that sleep maintains and enhances synaptic efficiency (Kevenau, 1997). The SW–REM sleep cycle, it is suggested, evolved to allow the repetitive activation of neural circuits in the brain while it is effectively isolated from the environment. The different EEG rhythms of the sleep stages represent the stimulation of neural circuits in different brain structures. The reason REM sleep is so prominent in the fetus is because that is when most of these circuits are being formed.

Learning theories propose that REM sleep permits, or at least promotes, the formation of long-term memories. Lots of evidence apparently demonstrates the role of sleep in learning and memory (Maquet, 2001). For example:

- post-training sleep deprivation impairs subsequent performance in animals and humans;
- there is increased REM on post-training nights, returning to normal when the task has been mastered;
- specific neural activity generated during training recurs during subsequent sleep (suggesting enhancement of synaptic connections);
- PET scans (see Topic A2) have shown that brain areas that were more active in people during a reaction time training task are also more active in subsequent REM sleep;
- in post-training sleep, different electrical rhythms can be recorded in cortical and subcortical areas involved in learning and in the consolidation of memory (see Topic P1). This suggests a two-stage process involving both REM and SW sleep.

This view accords with the observation that sleeping after learning, for example before an exam, helps us to remember what we have learned. However, this, and most of the other evidence, might be explained by other factors (Siegel, 2001). For example, sleep might reduce **retroactive interference** from later stimulation, or might reduce stress or tiredness that would impair performance. Furthermore, people taking the antidepressant drugs **MAOI** (see Topic R2), which eliminate REM sleep, for periods of months or years have unimpaired memory.

However, most workers accept that REM sleep, and probably also SW sleep, promotes the formation of memories. The two sleep stages may operate on different types of memory, or both might be involved sequentially in all memory consolidation. There are, though, limits to this:

- sleep influences the consolidation of **implicit or procedural memory** (memories of which we are not necessarily aware — e.g. a stimulus seen once will influence later recognition tasks; perceptual–motor skills), but not **explicit or declarative memory** (consciously learned facts, events or stimuli);
- the effects of sleep are limited to a narrow 'window' shortly after the acquisition of the response;
- sleep is not *essential* for learning, which of course takes place during waking, but does *enhance* it;
- memory consolidation is not the only function of sleep, as we saw above.

Dreams might simply be meaningless by-products of the sensory activation occurring during sleep, or they might be the conscious representations of the complex brain systems involved in the reprocessing of emotion and memory during sleep. Others have viewed dreams as the important component of sleep. Freud and his followers, of course, viewed dreams as a 'safe' way in which the individual could express (in a disguised manner) repressed drives. Sleep could be viewed as existing largely to serve this function. There is, however, little or no evidence to support this.

K3 MECHANISMS OF SLEEP AND AROUSAL

Key Notes

The reticular activating system	Cutting the brain stem above the pons produces a sleeping EEG, while transection below the brain stem does not affect the sleep–waking cycle. Small lesions in the reticular formation that leave many sensory pathways intact can also produce sleep, while destruction of the sensory pathways alone does not affect EEG patterns. Consciousness is not maintained by stimulation of the cerebral cortex by the reticular activating system.
Neural control of REM sleep	Centers in the brain stem control REM sleep. REM is initiated by activity in cholinergic neurons in the peribrachial area of the pons. Axons from there control the components of REM sleep. This is partly modulated by the locus coeruleus and the raphé nuclei that inhibit the cholinergic system.
Neural control of SW sleep	The preoptic area and the anterior hypothalamus (the POAH) in the basal forebrain are involved in controlling SW sleep. The POAH is close to the suprachiasmatic nucleus (SCN), enabling entrainment of sleeping and waking to a circadian cycle. Neurons in the POAH are involved in temperature regulation, providing a mechanism for the effects of body temperature on SW sleep.
Chemical mechanisms	Sleep does not result from the build-up of a sleep-inducing chemical during wakefulness. Some drugs produce insomnia, others induce sleep. Neurochemicals are involved in the neural mechanisms of sleep. The inhibitory neuromodulator adenosine accumulates in the basal forebrain and the POAH. Muramyl peptides produced by bacteria in the gut also seem to be involved in the production of SW sleep.
Related topics	The nature of sleep and dreams (K1) Biological rhythms (K4)

The reticular activating system

Brémer (1936) argued that consciousness is maintained by diffuse sensory input to the cortex, and that sleep results from *reduced sensory input*. Cutting through the brain stem above the pons in cats (producing the *cerveau isolé* preparation) resulted in an EEG typical of SW sleep. Transection below the brain stem (the *encéphale isolé*) did not disrupt the normal sleep–waking cycle (see *Fig. 1*). Brémer argued that the difference was that the higher lesion removed the sensory input to the cortex. Moruzzi and Magoun (1949) showed that stimulation of the **reticular formation**, located in the center of the brain stem between these two lesion sites, results in EEG arousal and behavioral excitation. This part of the reticular formation became known as the ascending **reticular activating system (RAS)**. At about the same time it was shown that smaller lesions in the upper part of the reticular formation produce the same effect as complete transection above the pons: SW sleep.

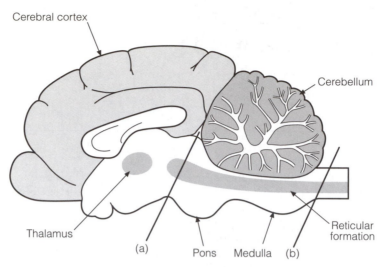

Fig. 1. Lesions of the brain stem in cats and their effects on sleep. (a) The cerveau isolé produces SW sleep. (b) The encéphale isolé has no effect on sleeping and waking.

These lesions leave many sensory pathways intact. Also, destruction of the sensory pathways but not the RAS does not affect EEG patterns. So the idea that sleep follows reduced sensory input was discarded.

Neural control of REM sleep

It was noticed in the late 1950s and early 1960s that anesthetizing a region in the brain stem, between the pons and the medulla (see *Fig. 1*) of a sleeping cat causes it to wake up. At about the same time evidence began to gather that the features of REM sleep could be separately affected by manipulations such as localized lesions and REM sleep deprivation (see Vertes, 1983). This led to the view that the brain stem contains centers in and around the reticular formation that actively control the onset and nature of REM sleep.

Cholinergic neurons in the **peribrachial area** of the pons pass axons to various brain areas. These neurons fire quickly during REM sleep, or during both REM sleep and waking (Steriade et al., 1990). Stimulating the sites to which these axons pass can produce components of REM sleep, while destroying the sites abolishes them. For example, the **medial pontine reticular formation** (**MPRF**) seems to be involved in the onset of all the REM sleep components. Direct projections to the **thalamus** control cortical arousal, and hence the desynchronized EEG, while those to the **lateral geniculate nucleus** of the thalamus produce PGO waves, and those to the **tectum** produce rapid eye movements. The inhibition of muscular activity is controlled by the **magnocellular nucleus** in the medulla, which is innervated indirectly from the peribrachial nucleus (Shouse and Siegel, 1992).

A **noradrenergic** system with its origins in the **locus coeruleus** in the pons also sends axons to many cortical and subcortical regions. The firing rate of cell bodies in the locus coeruleus has been shown to follow behavioral activity closely, dropping to near zero during REM sleep and rising rapidly when the animal wakes (Aston-Jones and Bloom, 1981). Similar findings relate to **serotonergic** neurons in the **raphé nuclei** (Trulson and Jacobs, 1979). This strongly suggests that the locus coeruleus and the raphé nuclei control the onset of REM sleep by inhibiting the cholinergic neurons in the peribrachial area. *Fig. 2* shows a simplified diagram of the neural centers mentioned here.

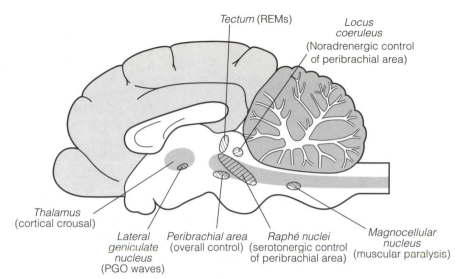

Fig. 2.   Some brain centers involved in REM sleep, in a cat's brain.

Neural control of SW sleep

Lesions of the **basal forebrain**, immediately in front of the hypothalamus, often cause animals to stop sleeping, although sleep may return after a few days (Trulson and Jacobs, 1979). While some evidence suggests that stimulation of this area produces SW sleep, others have not found this effect. Within the basal forebrain, the **preoptic area** and the adjacent **anterior hypothalamus** (together called the **POAH**) are close to the **suprachiasmatic nucleus (SCN)**. **Cholinergic** (and other) neurons in the basal forebrain have been found that are most active during waking. Axons from these pass diffusely to the cerebral cortex. It is likely that fibers from the SCN pass to the preoptic area to entrain sleeping and waking to a circadian cycle (see Topic K4). Neurons in the POAH are involved in temperature regulation, and receive inputs from temperature sensors in the skin. Local warming of this area produces SW sleep, and this provides a mechanism for the effects of body temperature on SW sleep (McGinty et al., 1994; see Topic K2). *Fig. 3* shows a simplified diagram of these neural centers.

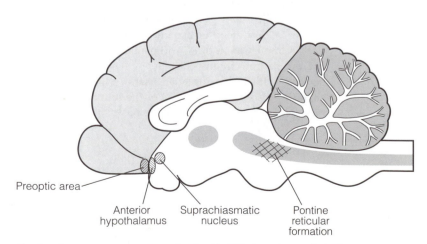

Fig. 3.   Some brain stem centers involved in SW sleep, in a cat's brain.

Chemical
mechanisms

One longstanding view of the cause of sleep is that it results from the gradual build-up of some sleep-inducing chemical, *'substance S'*, during wakefulness. Substance S would then be broken down during sleep, and when reduced enough would allow the animal to awaken. Piéron (1913), for example, claimed that the injection of cerebrospinal fluid from sleep-deprived dogs produced sleep in other dogs. A number of possible sleep-promoting substances have been identified over the years, including **melatonin**, secreted by the pineal gland. Melatonin levels rise during darkness (see Topic K4). However, it does not act as a sleep-promoting substance, except in very high doses.

One good reason for believing that circulating substances do *not* control sleep in humans comes from a study of conjoined twins who shared a circulatory system. Their sleeping and waking periods were found not to be coordinated. A circulating sleep-promoting substance also cannot explain those cases where animals sleep with only one cerebral hemisphere (see Topic K1).

An alternative view is that substances produced within the brain control sleeping and waking. It is certainly true that certain classes of drugs affect sleep. Some anti-inflammatory drugs (e.g. aspirin) can interfere with sleep by lowering body temperature. Others, such as the **benzodiazepines**, promote sleep, and are used to treat insomnia (see Topic K5). These act on **GABA** receptors, widely distributed in the brain, increasing the inhibitory effects of GABA. However, a naturally occurring brain chemical that this effect might mimic has not been discovered. Chemical involvement in sleep and waking is, of course, apparent in the neurochemicals that are involved in the neural mechanisms. We discussed mechanisms based on noradrenaline, acetylcholine and serotonin earlier. Another such substance is the inhibitory neuromodulator **adenosine**. Recent research suggests that adenosine accumulates in the basal forebrain and the POAH during waking. In the basal forebrain it directly inhibits the neurons that modulate the desynchronization of cortical activity. In the POAH it indirectly stimulates those that promote sleep (Strecker et al., 2000).

Muramyl peptides (**MPs**) are another chemical factor that might be involved in the control of sleep (Brown, 1995). These are substances produced not by the body itself, but by bacteria living normally in the gut. Removal of the bacteria reduces total sleep time in humans (no measures of sleep architecture were made, so we do not know what specific components were reduced), and SW sleep in rats. The fact that very little SW sleep occurs in neonates (see Topic K1) coincides with there being very few bacteria in the gut of neonates. While the absence of SW sleep is usually attributed to the immaturity of the neonatal brain, administration of MPs to neonatal rabbits increases SW sleep. This shows that the neural circuits exist but are not activated until the amount of MPs increases with increasing gut bacteria. MPs are known to be important factors enhancing various components of the immune system. As we saw in Topic K2, immune system modulation is one postulated function of sleep.

K4 BIOLOGICAL RHYTHMS

Key Notes

Circadian rhythms	The sleep–waking cycle represents the daily (circadian) variations in hormone secretion, metabolic activity and cognitive capacities. In people isolated from the normal cues of the day–night cycle this cycle advances about one hour each day. This demonstrates that there is a biological clock, entrained by external events (zeitgebers) to 24 hours.
Other biological rhythms	Ultradian rhythms include the basic rest–activity cycle with a period of about 90 minutes. It is seen in the different stages of sleep. Infradian rhythms include the menstrual cycle. Circannual rhythms are seen in hibernation and in animals with an annual breeding cycle.
Mechanisms of circadian rhythms	The cyclicity of the suprachiasmatic nuclei (SCN) of the hypothalamus is based on self-limiting protein synthesis. The SCN receive information from the optic chiasma, allowing entrainment of rhythms by light. The SCN exert neural and chemical control. There are other clocks controlling circadian rhythms for a wide range of bodily functions. The role of the SCN clock might be to synchronize all these individual rhythms.
Mechanisms of other rhythms	The SCN clock helps control infradian rhythms, causing the pineal gland to secrete melatonin at night. Melatonin feeds back to the SCN, and affects other brain centers controlling seasonal processes. Melatonin facilitates recovery from jet lag. Ultradian rhythms also seem to be controlled by endogenous clocks, influenced by periodic variations in homeostatic behavior. Lesions in parts of the basal hypothalamus, but not the SCN, can disrupt the basic rest–activity cycle.
Related topics	The nature of sleep and dreams (K1) Disruptions of sleep and rhythms (K5)

Circadian rhythms

The most obvious rhythm affecting us is the daily sleep–waking cycle. This is known as a **circadian rhythm** (meaning 'about daily'). Humans (and most other primates) are **diurnal** (meaning active in daylight), while many other animals (e.g. rodents) are **nocturnal**. Underlying the sleep–waking cycle are continuous variations in hormone secretion and metabolic activity, including a daily variation in body temperature of about 2°C. This coincides with continuous variation in cognitive capacities (e.g. ability to concentrate), which are highest when body temperature is highest: for us normally mid-afternoon. These cycles permit each animal to be most alert at times when it is adapted to be active.

When animals or people are kept in isolation from the normal cues of the day–night cycle, they wake up and go to sleep a little later each day (Wever, 1979; see *Fig. 1*). This is known as a **free-running rhythm**, and in humans usually has a period of about 25 hours. This demonstrates that

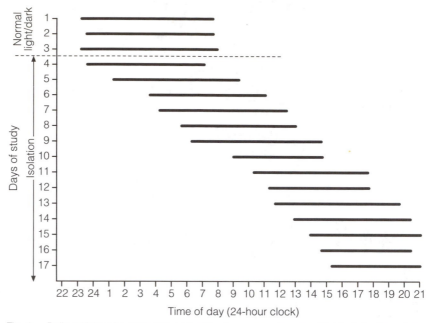

Fig. 1.   Daily activity of a man with and without exposure to normal light–dark alternation, showing a free-running cycle of just over 25 hours. Each black bar shows the time he was asleep each day.

- there is an internal (endogenous) mechanism underlying the circadian rhythm – some sort of **biological clock** with a cycle of about 25 hours;
- external events (called **zeitgebers**, from the German for time-givers) can *entrain* the circadian rhythm, normally keeping it to 24 hours.

The most important zeitgeber is light. Other zeitgebers have been demonstrated in hamsters, e.g. social interaction, feeding and exercise (e.g. Mistleberger, 1994). Humans use a variety of cues to help entrain the rhythm, including social interaction, feeding and alarm clocks.

Other biological rhythms

Ultradian rhythms occur more often than once per day (that is, they have a period less than 24 hours). The most important is the **basic rest–activity cycle (BRAC)**, with a period of about 90 minutes. Kleitman (1961) first described this in the demand feeding cycles of newborn infants. It is seen clearly in the way we cycle through different stages of sleep (see Topic K1). It also affects us during waking as variations in such things as attention, eating, drinking, heart rate, oxygen consumption, muscle tone and gastric motility. Its effects are not obvious, usually being masked by changes in activity and by variations in motivation.

Infradian rhythms ('less than daily') have a period longer than 24 hours. The most obvious human example is the menstrual cycle, with a period of about 27 days. Many other animals have **estrus cycles** of various lengths (see Topic N1). Another widespread type of rhythm is **circannual rhythms**. These are clear in hibernating animals, and in the behavior of animals, such as birds, with an annual breeding cycle. There is now some evidence of these in humans, particularly annual cycles of a depressive illness known as **seasonal affective disorder (SAD**; see Topic R2).

Mechanisms of circadian rhythms

Moore and Eichler (1972) and Stephan and Zucker (1972) discovered that lesions in a small area at the base of the medial hypothalamus, the **suprachias-matic nuclei (SCN)**, abolish rats' circadian rhythms. The SCN are two areas about the size of a pinhead immediately above the optic chiasma, where the optic nerves from the two eyes come together (see Topic G1). Although oper-ated rats sleep the same total amount as before, their sleep is distributed throughout the day and night (see *Fig. 2*). The SCN lesions affect the *cyclicity* of sleep, not the *need* for sleep.

The SCN are the *source* of the rhythmical activity. The metabolic rate of the cells in this region is periodic (Schwartz and Gainer, 1977). Electrical recordings from isolated pieces of SCN tissue, even single neurons, continue to show rhythmic activity synchronized to the light–dark cycle (Welsh et al., 1995). The origin of the rhythm in the cells is in the synthesis of proteins called **PER** and **TIM**, which accumulate in the nuclei of cells in the SCN until they reach a concentration that switches off their own production. The genes that produce these proteins have been identified in fruit flies and mammals, as have genes coding for proteins that regulate PER and TIM gene expression (e.g. Tei et al., 1997). The synthesis of the proteins involved is directly affected by light, becoming active when darkness falls.

The SCN receive information directly from the optic chiasma, and hence from the retina. Lesion studies and stimulation studies have shown that this affects the entrainment of circadian rhythms. This is clearly the route through which the primary zeitgeber, light, modulates the activity of the SCN. The SCN are connected neurally to many other parts of the brain that control functions following a circadian rhythm. However, SCN neurons transplanted into animals that have previously had their SCN destroyed start to have an effect before neural connections are established. This shows that some of the control of the SCN is exerted through chemical rather than neural processes.

There is evidence for more than one clock controlling circadian rhythms. If the isolation in a free-running experiment continues for more than about 30

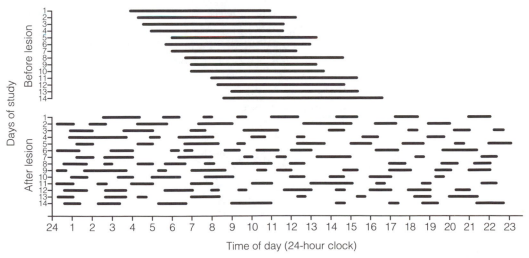

Fig. 2.   Simplified diagram of sleep in a free-running rat before and after bilateral destruction of the SCN. Before the lesion sleep is concentrated in one main bout each day. After the lesion it is distributed throughout each 24-hour period.

days, there is a **desynchronization** of the sleep–waking and temperature cycles in many people. The sleep–waking cycle lengthens to 30 hours or more, and the temperature cycle remains at about 25 hours (Aschoff, 1994). Animals presented with food at the same time each day come to anticipate it, shown by increased activity, even in isolation from environmental cues. This must depend on some internal clock, and it persists if the SCN are destroyed. Recently, it has been discovered that many other processes throughout the body are rhythmic. At least 160 genes in the fruit fly show daily variations in gene expression. These cyclic genes code for proteins involved in a wide range of processes, ranging from memory to locomotion. It is also known that the effects of many drugs, for example some used in cancer treatment, vary according to the time of day when they are administered. At least part of this is due to cyclic variations in response of peripheral blood vessels to hormones and other circulating substances. Quite possibly, the role of the SCN clock is to synchronize all these individual rhythms (Moore-Ede, 1992).

Mechanisms of other rhythms

The clock in the SCN also helps to control *infradian* rhythms. For example, many animals have an annual breeding season that commences when the day length starts to increase. Males then start to secrete more testosterone. Destruction of the SCN often abolishes this *circannual* cycle, and animals produce constant amounts of testosterone throughout the year. The SCN have neural connections with the **pineal gland**, and cause the pineal gland to secrete **melatonin** at night. Melatonin in turn affects the SCN, and also other brain centers controlling seasonally variable processes. Longer nights result in a greater concentration of melatonin, so that seasonal changes affect body cycles through melatonin (Bartness et al., 1993). While melatonin normally has very little effect on adult humans (removal of the pineal gland in adults has not been shown to have any effect), it is crucial for the seasonal variations in other mammals. Administration of melatonin to humans has, however, been found to facilitate recovery from jet lag (see Topic K5).

Ultradian rhythms also seem to be controlled by an endogenous clock, or clocks, although they are greatly influenced by variations in homeostatic behavior such as searching for food. SCN lesions that abolish the circadian rhythms do not necessarily affect the BRAC. However, lesions in other parts of the basal hypothalamus can disrupt the BRAC, although it is not firmly established whether these are the locations of a clock or clocks, or pathways through which a mechanism elsewhere affects behavior (Gerkema and Daan, 1985). The midbrain mechanisms controlling the onset of REM sleep could be viewed as the 'clock' controlling the BRAC (see Topic K3).

K5 DISRUPTIONS OF SLEEP AND RHYTHMS

Key Notes

Insomnia	Insomnia often results from emotional arousal. Benzodiazepines, used to treat insomnia, have little actual effect on sleep, and stopping taking them causes further insomnia. Sleep apnea is when a sleeping person stops breathing, and is awoken when the rising level of carbon dioxide stimulates central chemoreceptors.
SW sleep disorders	Sleepwalking, enuresis or bed-wetting, and night terrors occur in early childhood. They usually disappear without treatment.
REM sleep disorders	In narcolepsy, sufferers suddenly fall asleep for a few minutes during the day, and may collapse with the accompanying loss of muscle tone (cataplexy). Narcolepsy can be treated with tricyclic antidepressants. In REM sleep behavior disorder there is an absence of the usual atonia of REM sleep. Patients sometimes commit violent acts during these events.
Disruption of circadian rhythms	Jet lag is the sleep and mood disturbance and loss of concentration resulting from our internal clocks suddenly becoming out of synchrony with the local zeitgebers. A similar result occurs when changing work shifts. Recovery follows resynchronization of the internal rhythm and the local environment. Jet lag can be minimized by gradually shifting the waking up time to earlier in the morning, exposure to intense light early in the morning, or treatment with melatonin.
Related topics	The nature of sleep and dreams (K1) Biological rhythms (K4)

Insomnia

Insomnia, which is lack of sleep or difficulty falling asleep, is reported by 20–30% of the population. Insomnia is defined subjectively. People have been demonstrated to function well and happily on as little as 1 hour of sleep each day. Many of those who report that they sleep little, and that they do not sleep enough, are actually found when observed to sleep much more than they estimate. People describe themselves as insomniac because they feel or believe that they should sleep more than they do. In some of these people, daytime behavior and performance are affected. Insomnia often results from emotional arousal. For example, anticipation of both positive and negative upcoming events can disturb sleep. Clinically, anxiety is often associated with difficulty falling asleep (see Topic R3), and depression with frequent or early awakening (see Topic R2).

Persistent insomnia is often treated with drugs that are also used to treat anxiety. The most commonly used are the **benzodiazepines**, including Valium and Librium. Unfortunately, users develop tolerance to these drugs, resulting in the need to increase the dose. More importantly, stopping taking a drug to which one has developed tolerance leads to a rebound effect (see Topic F4); in this case, taking sleeping drugs *causes* insomnia. Furthermore, studies of people taking benzodiazepines have shown that their effect is minimal. In one such study, the drug only reduced the average time to fall asleep from 30 minutes to 15 minutes, and increased the average sleep time by only half an hour.

A particular type of insomnia is **sleep apnea**. In this, the sleeping person stops breathing, and is awoken when the rising level of carbon dioxide in the blood stimulates central chemoreceptors. The sufferer does not necessarily recall this in the daytime, and may simply report tiredness. Sleep apnea seems to have two main causes:

- restriction of the airways (sleep apnea is particularly common in those who snore);
- failure of the CNS to initiate breathing at normal levels of carbon dioxide.

SW sleep disorders

Sleepwalking is not, as might be thought, the enactment of dreams, since it occurs during SW sleep, when narrative dreams are least likely (see Topic K1). Sleepwalking is most common in childhood, and, while the cause is not known, it does not indicate any underlying pathology.

Enuresis, or bed-wetting, is common in early childhood. Its persistence is usually due to the failure of the child to awaken in response to stimuli from the bladder. More rarely, it may indicate anxiety-based conditions. In most cases it can be treated with a 'pad-and-bell' device, in which a moisture-sensitive pad is placed under the sheets. The first drops of urine complete an electrical circuit causing a bell to ring, wakening the child. The child quickly learns to awaken in response to the bladder stimuli before bed-wetting occurs.

Night terrors also occur most often in children. The child wakens suddenly, terrified, from SW sleep. They are not waking from a nightmare (which is a frightening dream during REM sleep). Night terrors generally disappear with age without any treatment.

[handwritten margin note: often found in children]

REM sleep disorders

In **narcolepsy**, sufferers are suddenly overcome by an irresistible urge to fall asleep during waking hours. While it may happen at any time, it most often happens in monotonous situations. The sleep typically lasts for 5–10 minutes, and the individual wakens feeling refreshed. Narcoleptics often pass straight into REM sleep on falling asleep; unlike the normal sleep pattern (see Topic K1). On other occasions, the narcoleptic person may suddenly lose all muscle tone and will collapse, usually fully conscious. This is known as **cataplexy**. The paralysis seems to be the same as that normally occurring during REM sleep. A strain of dog has been bred that suffers from narcolepsy. This indicates that narcolepsy can be a genetic disorder, and the gene causing it in dogs has been discovered (see Aldrich, 1992). However, the genetic basis appears to be more complex in humans. The source of the problem seems to be an excessive excitability of cholinergic neurons in the peribrachial area, stimulating the cells in the magnocellular nucleus that produce the muscular paralysis of REM sleep (Nishino et al., 1994; see Topic K3). Narcolepsy can be treated with tricyclic antidepressants, one effect of which is to reduce REM sleep.

REM sleep behavior disorder (**RSBD**), sometimes called adult sleepwalking, is distinct from childhood sleepwalking. Sufferers do not show the atonia usual during REM sleep, and may act out their dreams, often becoming violent (Mahowald et al., 1990). It is likely that RSBD involves interference with the mechanism in the magnocelluar nucleus in the medulla that normally produces the atonus of REM sleep (see Topic K3). However, the causes of this are not clear, and may differ in different cases (Ferini-Strambi and Zucconi, 2000).

Disruption of circadian rhythms

Flying across time zones, or changing between day- and night-working shifts causes our internal clocks to become suddenly out of synchrony with the local zeitgebers. Flying east across the Atlantic Ocean, for example, results in our peak alertness etc. being 6 or 7 hours later than local clock time. This is a **phase advance**; the zeitgebers are in advance of the previously established cycle. The result is sleep and mood disturbance and loss of concentration during local waking hours, which we call **jet lag**. In each case, our internal rhythm and the zeitgebers are out of synchrony. Recovery from jet lag follows resynchronization of the internal rhythm and the local environment (see *Fig. 1a*). Westward flight (**phase delay**) results in less jet lag than eastward flight, and we adjust to it more rapidly. This is because eastward flight results in a shorter night during the transition, and in order to synchronize with the new zeitgebers, we have to try to fall asleep *later* when we have gone west, but *earlier* when we have flown east. It is easier to delay sleep than to go to sleep earlier. Adjusting fully to an eastward flight might be expected to take about one day for every hour of time shift. Similarly, if we move to a later work shift we suffer less than if we move to an earlier shift.

The effects of jet lag can be minimized by gradually shifting the waking up time to earlier in the morning, which delays the circadian rhythm, accelerating entrainment. Exposure to intense light early in the morning, or a burst of intense exercise, both facilitate rapid resynchronization (Boulos *et al.*, 1995; see *Fig. 1b*). Treatment with the hormone **melatonin** produced by the pineal gland also promotes adjustment to such phase shifts (Deacon and Arendt, 1996).

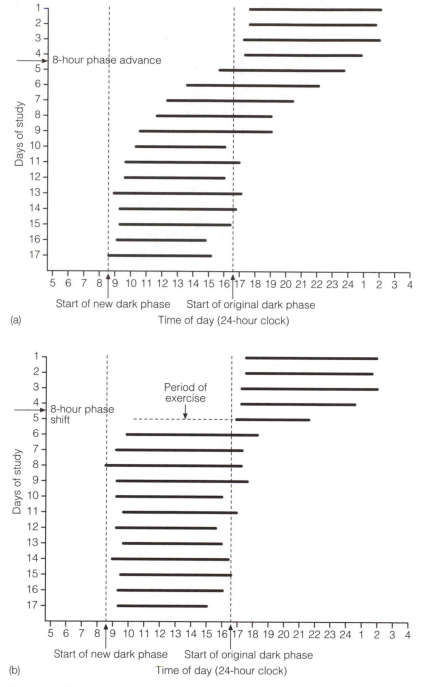

Fig. 1.   (a) Sleep periods in a hamster before and after a sudden advance of 8 hours in the light–dark cycle. (b) Faster synchronization follows a period of intense exercise.

L1 HOMEOSTASIS

Key Notes

Homeostasis	Homeostasis refers to the physiological processes that maintain the optimal stability of the internal environment.
Negative feedback systems	Homeostasis operates through negative feedback. Control systems in the body are much more complex than simple systems such as a thermostat. Homeostasis needs to control the system variable within a very narrow range around a set point. Values of each system variable are detected by a number of sensors. The values are compared with the set point, and if they are different, controls start bidirectional correctional processes. Physiological control systems have high levels of redundancy in all components of the feedback system.
Homeostatic behavior	All animals use behavioral means to assist homeostasis. These behaviors are called homeostatic behavior. Ectotherms such as reptiles have only behavioral means of controlling body temperature. Endotherms, including humans, use behavioral and physiological ways of controlling temperature.
Temperature regulation	The main means endotherms have of generating heat is metabolism; the release of energy in the tissues, especially muscles. Brown adipose tissue in the abdomen generates heat by increased metabolism. Most heat is lost from the body by evaporation of water in sweat and from the lungs, and through convection and radiation from the skin. Neurons in the preoptic area and the anterior hypothalamus change their rate of firing in response to changes in brain temperature. Further sensors are found in the skin, and there are other control centers in the midbrain and spinal regions.
Related topics	Thirst and drinking (L2) Physiological mechanisms in eating (M2)

Homeostasis

Animal cells and tissues work optimally only when their operating conditions are kept within a very narrow range. There are mechanisms for controlling every aspect of the internal environment to maintain these optimal conditions, such as temperature, electrolyte concentrations, pH (acidity) of body fluids, oxygen level, carbohydrate concentrations of tissues, and so on. These conditions have to be maintained in the face of fluctuations in the demands that the environment makes on the body. The physiological process that produces this stability is known as **homeostasis**, a term introduced by Cannon (1932). Homeostasis operates through **negative feedback**.

Negative feedback systems

The operation of a negative feedback system is often compared with a thermo-statically controlled heating system. As we shall see, the analogy is a poor one, although it illustrates the basic principles of negative feedback control, and allows us to recognize the limitations of a simple system. *Fig. 1* shows the essential features of such a control system. These are:

- the **system variable**; the property to be controlled. In the case of a thermostat, this is room temperature;
- the **set point**; the target value of the property, in the example a temperature that the system tries to maintain;
- a **sensor**, to detect and report the current state of the system. In this example this would be some form of thermometer;
- a **comparator**, which compares the output from the sensor with the set point, or which detects change in the system variable to test if the system variable is different from the set point;
- a **control**, such as a switch, which is a mechanism to start and stop the last component;
- a **correctional process**; in a thermostatic system, a heater.

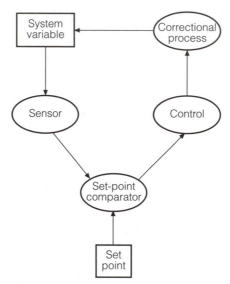

Fig. 1.   The components of a simple negative feedback system.

Frequently, the sensor, the set-point comparator and the control are combined. For example, in a room thermostat, the sensor might be a bi-metallic strip that bends when the temperature rises and falls, essentially registering the change and switching a heater on and off.

Such a control system is described as a negative feedback system because *increases* in the system variable feed back to the control to *reduce* the correctional process, and vice versa. In the case of a room heater, when the sensor detects that the temperature has fallen below the set point, it causes a switch to start a heater, which causes the temperature to rise. When the thermometer detects that the temperature has risen above the set point, it causes the switch to turn off the heater.

While a simple system like a room thermostat might be adequate for controlling room temperature, it has weaknesses that could be fatal in a homeostatic mechanism. First, a thermostat cannot maintain an exact temperature. Instead, the temperature at which it switches the heater on is always lower than the temperature at which it switches it off. Usually, this difference is quite large, as you can tell for yourself by turning the control on a room thermostat up and then down. You can hear the click of the switch turning off and on at different places on the thermostat. In a physiological system this would lead to a dangerously wide variation in internal conditions. We can improve on this by having a heater with continuously variable output, controlled by detected variations in temperature. Such a system is a **servo** system, and it retains the essential characteristic of working on the basis of negative feedback. Physiological control systems are more like servos.

A second limitation is that a simple thermostatically controlled heating system permits correction only in one direction. If the temperature rises above the set point, there is no way to lower it. This can be overcome by adding a second correctional process that would cool the room if it gets too hot. We can do this with air conditioning, which again operates as a negative feedback system. This bi-directional control is also a feature of physiological systems.

Another difficulty with the room heater analogy is that it is vulnerable to failure of components or of the connections between them. If any one of these fails, the whole system fails. The solution to this is to build **redundancy** into the control system. This means having more than one of each type of component, preferably with more than one connection between components. It would also be an advantage to have correctional processes of different types, which would allow the system to cope if, for example, one connection failed. For similar reasons, we might expect to find that there is more than one means of stopping the correctional process, for example cutting the fuel supply or diverting the heat elsewhere. It would also be an advantage to have such 'off switches', which in physiological systems are known as **satiety mechanisms**, separate from the 'on switches'. These are all characteristic of homeostatic systems.

Homeostatic behavior

Mammals, birds and other **endotherms** are able to maintain their body temperature within a very narrow range that is optimal for biochemical activity, and hence physiological processes. Other groups of animals, **ectotherms** such as reptiles and amphibians, cannot control their own temperature by internal mechanisms. Ectotherms slow down when the environment gets colder, and speed up when it gets warmer. The only ways that they have of controlling temperature are behavioral. To increase body temperature, they seek sunlight, and orient their bodies to maximize the absorption of heat. To cool themselves, they seek shade. Mammals also engage in such behavior (for example, they usually prefer to rest in warmer parts of the environment). We refer to behavior like this as **homeostatic behavior**. The control of temperature and, we will see later, other physiological states, is accomplished by the integrated actions of physiological homeostatic mechanisms and homeostatic behavior.

Temperature regulation

As the previous section suggests, endotherms such as humans have mechanisms for generating and dissipating heat. The main means we have of generating heat is **metabolism**; the release of energy by the breaking down in the tissues of chemical substances (see Topic M1). The more active we are, the more heat we produce, principally in the muscles. When we are cold, muscles may go involuntarily into asynchronous contractions, *shivering*, to generate heat. In the abdomen are deposits of **brown adipose tissue** (brown fat). This tissue is adapted to generate heat by increased metabolism, and is stimulated to do so by the sympathetic nervous system when the body temperature drops.

Heat is lost from the body by evaporation of sweat and of water exhaled from the lungs as water vapor. More heat is lost through convection and radiation from the skin. These cooling mechanisms can be modulated by, for example, increasing and decreasing the amount of blood flowing through small blood vessels in the skin, or by increasing or decreasing sweat gland activity. Small animals have a large body surface area in relation to body weight, and so lose heat more rapidly than large animals. They therefore have to generate heat more rapidly, and so have a higher metabolic rate and eat more food per unit of weight. Fur and feathers are adaptations that reduce heat loss through the skin. Mammals and birds can fluff up their fur or feathers to increase the insulation.

All of the above are *correctional processes*. There are *sensors* located in the central nervous system. Neurons distributed in the **preoptic area** and the **anterior hypothalamus** change their rate of firing in response to changes in brain temperature. Further sensors are found in the skin, giving a faster response to environmental temperature. There are also multiple *controls*. Lesions in the **lateral hypothalamus** disrupt behavioral correctional processes, but not autonomic ones, while the reverse is true of lesions in the preoptic area. Further, there appear to be other control centers located in the midbrain and spinal regions. These seem to permit increasing variation from the set point; that is, they do not control temperature so closely. Although based on different mechanisms and pathways, these centers normally operate apparently as one system.

L2 THIRST AND DRINKING

Key Notes

Fluid compartments

Water in the body is contained in fluid compartments: intracellular fluid and extracellular fluid. The latter comprises: interstitial fluid, blood plasma and cerebrospinal fluid. Water balance is necessary to maintain intracellular electrolyte concentrations, and the total volume of water. Water passes through semipermeable membranes of the cell walls from regions of lower concentration to higher concentration (osmosis).

Regulation of fluid balance

Water and electrolytes are lost through processes including sweating, urination, defecation and bleeding. Water alone is lost by breathing and evaporation. The conscious sensation of thirst occurs if the body reaches a state of deficit.

Control mechanisms

Hypovolemic thirst follows loss of both water and electrolytes. Blood flow detectors in the kidneys cause the production of renin, which leads to the formation of angiotensin II. This causes constriction of peripheral blood vessels, stimulates the secretion of aldosterone (which causes the kidneys to reabsorb more sodium), and acts on the subfornical organ, stimulating drinking. Baroreceptors in the heart and major blood vessels stimulate production of antidiuretic hormone, causing the kidneys to retain water. Osmometric thirst results from stimulation of osmoreceptors in the organum vasculosum of the lamina terminalis and the supraoptic nucleus of the hypothalamus. The median preoptic nucleus seems to be the common pathway that initiates drinking.

Satiety mechanisms

We drink about the right amount of water to correct any deficit, although it takes several minutes for ingested water to rehydrate the extracellular fluid. Wetting the mouth temporarily quenches thirst. This is a short-term satiety mechanism. The long-term mechanism is based on the osmoreceptors in the throat, in later parts of the gastrointestinal tract and in the liver. These inhibit ADH secretion when stimulated by water.

Food-related drinking

Usually we drink before a deficit has occurred. We drink during or after eating: food-related drinking. This is mediated by angiotensin, histamine and insulin.

Spontaneous drinking

Drinking in the absence of deficits is motivated by the positive incentive properties of drinking. With increasing deprivation preference shifts towards drinks with less flavor and lower concentrations of salt and sugars. Consumption is influenced by other sensory qualities of the drink, as well as by cultural and social factors. We learn to drink to avoid deficits.

Related topics

Homeostasis (L1)

Physiological mechanisms in eating (M2)

Fluid compartments

Water in the body is contained in various **fluid compartments** (see *Fig. 1*). About two-thirds of the water is contained inside cells: the **intracellular fluid**. The remaining third is the **extracellular fluid**. This in turn has three components: the **interstitial fluid**, which separates cells from one another, the **blood plasma**, and the **cerebrospinal fluid** (see Topic C1).

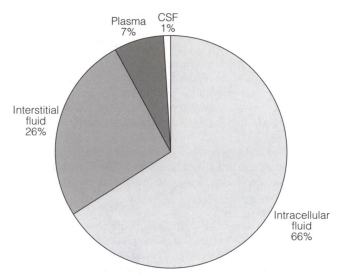

Fig. 1.   The fluid compartments of the body.

The amount of water in the body needs to be controlled for two main reasons. First, cellular processes depend on chemical reactions between substances in solution within the cells. While most of these reactions take place in specialized structures in the cells, they depend on the presence of just the right amount of water in the cell. If water is lost from the intracellular fluid, the cell goes outside the optimal conditions for cellular processes. The intracellular fluid is separated from the interstitial fluid by complex cell membranes, **semipermeable membranes**, which contain receptors for various chemical substances (see Topic E1). These strongly resist the passage of many inorganic ions such as sodium, while freely permitting the passage of water molecules and waste products of cellular metabolism. Normally, the intracellular and interstitial fluids are *isotonic*. That is, the fluids maintain a balance of concentration of dissolved substances (solutes) inside and outside the cells. If the concentration of a solute inside a cell rises (it becomes *hypertonic*), water will pass into the cell to re-establish the equilibrium. If the concentration within the cell falls (becoming *hypotonic*), water will pass out of the cell until the intracellular and interstitial fluids are again isotonic. This process is called **osmosis** (see *Fig. 2*).

In addition to permitting the passage of food, waste, and other substances such as some hormones into and out of cells, the interstitial fluid acts as a **buffer**. It allows the immediate correction of the electrolyte concentration of the cells, so that the essential biochemical processes may continue. The interstitial fluid is in turn in contact, via semipermeable membranes, with blood plasma in the capillaries. In order for ingested water to influence the intracellular fluid, it has to be absorbed from the gastrointestinal tract into the bloodstream, from there into the interstitial fluid, and thence into the cells.

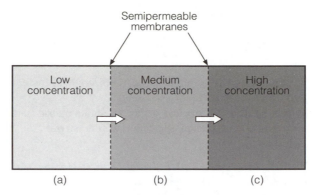

Fig. 2.   Osmosis: (a) is hypotonic with respect to (b), which is hypotonic with respect to (c). Water passes across the semipermeable membranes as shown by the arrows.

The second reason to control the water content of the body is to maintain the total *volume* of water. This allows blood pressure to be maintained, which is essential for proper circulation of blood to all tissues.

Regulation of fluid balance

Many of the things our bodies do, including sweating, breathing and the excretion of urine, involve a net loss of water from the body. Some of these processes lose only water (e.g. breathing), while others (e.g. sweating) lose electrolytes, particularly sodium and chloride ions, as well as water. Clearly, the control of water in the body must be closely connected with the control of electrolytes, and particularly of sodium. We normally ingest more water and more sodium ions (mostly from common salt) than we need. Most water and excess sodium are lost through urination. Significant amounts are lost through sweating, and through hemorrhage, both following trauma and in menstruation. The amount lost through sweating rises enormously during exercise, and also when the environmental or bodily temperature rises. The major loss of water without loss of electrolytes is evaporation, mainly through breathing but also through the skin. Water is also lost through defecation.

The conscious sensation of **thirst**, the feeling of a need to drink water, only occurs if the body reaches a state of deficit. Thirst is usually experienced after such things as exercise, eating common salt (e.g. in salted peanuts) or taking a diuretic substance such as alcohol. It also occurs after sudden blood loss, but only in severe cases; blood donors, who lose about 10% of their blood volume, do not usually report thirst.

Control mechanisms

Two interlocking control mechanisms have been distinguished. When we lose both water and electrolytes, the effect is a loss of fluid volume, resulting in **hypovolemic** (or **volumetric**) **thirst**. The existence of this as a separate system is demonstrated by taking blood or by the injection into the peritoneal cavity (in the abdomen) of substances (colloids) that draw plasma from the blood. Both of these cause a rapid decrease in the volume of the extracellular fluid. Neither of these changes the electrolyte balance of the body, yet each results in increased drinking (Fitzsimmons, 1961). The change in volume is detected by blood flow detectors in the kidneys, and by pressure receptors (**baroreceptors**) in the atria of the heart and in major blood vessels. In response to a fall in blood flow, the cells in the kidneys secrete the enzyme **renin**, which enters the bloodstream and causes the formation of the hormone **angiotensin II**. This has three main actions:

- it causes constriction of peripheral blood vessels (immediately increasing blood pressure);
- it stimulates the adrenal cortex to secrete **aldosterone**, which causes the kidneys to reabsorb more sodium;
- it acts on the **subfornical organ** (a small neural center close to the lateral ventricles, and outside the blood–brain barrier; see Topic C1), stimulating drinking.

The baroreceptors detect a decrease in stretching of the walls of the heart or blood vessels. This mechanism is revealed by several experiments. Experimentally reducing the blood flow to the heart, by inflating a tiny balloon in the vena cava, stimulates drinking in dogs (Fitzsimmons and Moore-Gillon, 1980). Cutting the nerve fibers from the region of the heart containing the baroreceptors prevents this effect. Chemical blocking of angiotensin receptors does not affect such results, so the effect is independent of renin production by the kidneys. The baroreceptors stimulate production of **antidiuretic hormone** (**ADH**) by the **posterior pituitary gland**. ADH causes the kidneys to retain water, and also stimulates the production of renin (see above).

The second type of thirst, **osmometric thirst**, results from stimulation of **osmoreceptors**. These are neurons sensitive to cellular dehydration. They are stimulated when water leaves the cells in response to loss of water and not electrolytes from the body, or to an increase in the amount of sodium in the body (e.g. from eating salty food). Osmoreceptors are located in the **organum vasculosum of the lamina terminalis** (**OVLT**), close to the third ventricle, and the adjacent **supraoptic nucleus** of the hypothalamus (McKinley, Pennington and Oldfield, 1996). Further osmoreceptors in the throat, the stomach and the liver do not result in immediate drinking. They operate through a different feedback loop that releases ADH, which, as noted above, causes water retention in the kidneys.

Although all the pathways are as yet uncertain, the OVLT osmoreceptors and the subfornical organ are connected to the nearby **median preoptic nucleus** (**MPN**; see Lind and Johnson, 1982). The MPN also receives direct sensory inputs from the baroreceptors in the atria of the heart. This center seems to be the common pathway that initiates drinking by way of behavioral circuits in the **periaqueductal gray** (**PAG**) area of the brain stem (see *Fig. 3*).

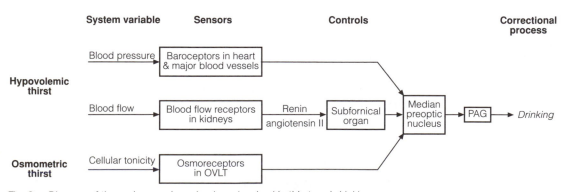

Fig. 3.   Diagram of the main neural mechanisms involved in thirst and drinking.

Satiety mechanisms

The mechanisms described above cannot act alone as satiety mechanisms, since there would be a long delay before they could change the situation sufficiently to stop ingestion. We generally drink just about the right amount of water to correct any deficit, although it takes several minutes for ingested water to rehydrate the extracellular fluid (Ramsay, Rolls and Wood, 1977). The obvious satiety mechanism would be osmoreceptors in the mouth. Wetting the mouth does quench our thirst, but only temporarily. Rats have been surgically given fistulas to carry water out of the esophagus before it reaches the stomach, producing **sham drinking**. If operated rats are then made thirsty, they cease drinking briefly immediately they wet their mouths. However, they quickly resume, and drink copiously. This brief cessation of drinking does not occur if the same quantity of water is introduced through a tube directly into the stomach.

This suggests that there are two satiety mechanisms. One is a short-term mechanism, which is strongest in the mouth and weakest further down the gastrointestinal tract. The other is a long-term mechanism, based on the osmoreceptors in the throat, in later parts of the gastrointestinal tract and in the liver. These inhibit ADH secretion when stimulated by water. Passing water into different levels of the gastrointestinal tract in animals has shown that the long-term inhibitory effect on drinking increases the further down the tract the water is introduced (see Verbalis, 1991).

Food-related drinking

These thirst mechanisms control **deficit-induced** drinking. But we do not drink simply in response to deficits; we drink before a deficit has occurred. One instance is the anticipation of the need for water resulting from eating: **food-related drinking**. Laboratory rats drink about two-thirds of their total water intake during eating and drink little if deprived of food. Kraly (1990) and colleagues have shown that drinking elicited by eating is mediated by central angiotensin receptors, as well as by peripheral and central histamine receptors. Histamine is produced by the stomach as a direct or conditioned response to food, and acts on the kidneys stimulating the production of renin, and hence angiotensin in the blood. Insulin secreted by the pancreas in response to eating (see Topic M1) is also involved in the control of food-related drinking.

Spontaneous drinking

We normally do not wait until we are deprived before drinking. People allowed to drink throughout a normal day report variations in thirst, dryness of the mouth, and pleasantness of the taste of water not associated with physiological variables. Drinking in the absence of deficits is motivated by the **positive incentive properties** of drinking; drinking is pleasant, and that is why we do it (Booth, 1991). Unless deprived, we prefer drinks that are flavored. With increasing deprivation our preference shifts towards drinks with less flavor and with lower concentrations of salt and sugars. On the other hand, studies with human beings under conditions leading to dehydration have shown that if the water available is made unpalatable, people will not drink enough to re-hydrate their tissues fully (Engell and Hirsch, 1991). Laboratory rats drink less if their food is made bitter.

Consumption is influenced by the temperature of the drink, so that any drink will have a particular temperature range that provides maximal motivation to drink it. This optimal temperature will depend on a variety of sensory factors (so that, for example, many white wines are preferred at low temperatures, and red wines at higher temperature), as well as cultural ones. Finally, drinking is affected by social factors, so that a person's desire to drink is higher when more people are drinking.

We learn to drink to avoid deficits. Laboratory rats can be conditioned to drink in response to a signal previously associated with an injection of formalin, which produces hypovolemia. Athletes regularly drink water or isotonic drinks while exercising, to maintain the fluid balance of their bodies and avoid deficits.

M1 DIGESTION, ENERGY USE AND STORAGE

Key Notes

Food	Each of the macronutrients (carbohydrates, proteins and fats) provides energy. Proteins provide amino acids. Lipids (fats), most vitamins, and all essential minerals must be ingested in food. Digestion breaks down food allowing it to be absorbed into the body. Digestion takes place in the gastrointestinal tract, comprising the mouth, esophagus, stomach, small intestine and large intestine. Complex carbohydrates are broken down into simple sugars, proteins into amino acids, and lipids into fatty acids and monoglycerides. Water and electrolytes are absorbed from the colon.
Energy use and storage	During the absorptive phase of metabolism nutrients are absorbed into the bloodstream, and most of the energy is stored. Excess carbohydrates are converted into lipids in the adipose tissues. Glucoreceptors in the liver cause insulin secretion, which permits all tissues to use glucose. In the fasting phase energy is released to supply the tissues. In the absence of insulin only neural tissues are able to metabolize glucose. When glucose levels fall the pancreas secretes glucagon, which causes the reconversion of glycogen to glucose, and converts triglycerides to fatty acids and glucose. In this phase fatty acids are the fuel for most tissues, while glucose remains the energy source for neural tissues.
Related topics	Physiological mechanisms in eating (M2) Weight control and its disorders (M4)

Food

Carbohydrates, proteins and fats are the three main food groups (**macronutrients**). Each provides energy. In addition, proteins provide the key building blocks of tissues, **amino acids**, some of which cannot be synthesized by the body. Fats (in general **lipids**) are important in many physiological processes. Some lipids, the *essential fatty acids* cannot be synthesized by the body. Most vitamins, and all of the minerals that are essential for the proper functioning of the body, must be ingested in food. Many macronutrients have molecules that are too large to be absorbed by the body and used by it. **Digestion** is the process of breaking down food, mechanically and chemically, into simpler substances allowing them to be absorbed from the small intestines and used by the cells of the body.

Digestion takes place in the **gastrointestinal** tract. The main components of this are the mouth, the esophagus (gullet), the stomach, the duodenum (the first 22 centimeters or so of the intestine), the small intestine (small in diameter, but 5–7 meters in length), and the large intestine (or *colon*; about 1–2 meters). Digestion starts in the mouth. Chewing reduces food to smaller pieces, and mixes it with saliva. Saliva lubricates the food, and adds an enzyme that helps the breakdown of the complex carbohydrate *starch* to simple sugars. The food is swallowed, and actively moved down the **esophagus** towards the **stomach** by rhythmic contractions called **peristalsis**. The food remains in the stomach, where it is mixed with hydrochloric acid, mucus, and further enzymes, notably **pepsin**, secreted by cells in the stomach lining. Contractions of the stomach help break the food down into smaller particles, and the acid helps the action of the digestive enzymes. Pepsin starts to break proteins down into their constituent amino acids. The stomach releases its contents gradually into the **duodenum**, where they are mixed with more enzymes, some from cells in the gut wall, and others from the **pancreas** and from the **liver** by way of the **gall bladder**. These assist the breakdown of complex carbohydrates into simple sugars, and proteins into amino acids. These simpler compounds are absorbed into the bloodstream through the wall of the duodenum and the rest of the small intestine. **Bile**, also secreted into the duodenum from the gall bladder, causes lipids to emulsify (to become tiny droplets suspended in water). Enzymes from the pancreas break these down into **free fatty acids** and **monoglycerides**, which are absorbed from the gut.

By the time the gut contents reach the **colon**, almost all of the nutrients have been absorbed. In the colon, water and electrolytes are absorbed, and the remaining material is passed from the anus as feces.

Energy use and storage

The tissues of the body obtain most of their energy from the metabolism of **free fatty acids** or of **glucose**. Although we need energy all the time, we only eat occasionally. For this reason, most of the energy in ingested food is stored. (*Fig. 1* summarizes the storage and release of energy in the body.) Metabolism is conventionally divided into three phases. The **cephalic phase** prepares the digestive system for the ingestion of food, and continues until the nutrients are about to be absorbed into the bloodstream. It starts not only in response to the

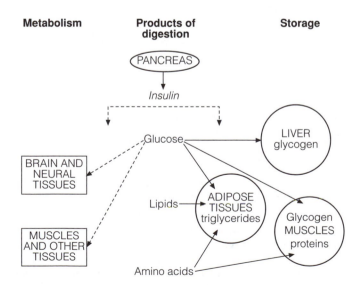

Fig. 1.   Energy storage and usage in the absorptive phase. In the presence of insulin all tissues metabolize available glucose. Excess glucose and other nutrients are stored as shown.

sight, smell and taste of food, but also to stimuli that have become associated with food (e.g. sounds of food preparation; Pavlov's famous demonstration of the conditioning of salivation to a sound paired with food presentation, Topic P2), and also with the *expectation* that food is due (see circadian feeding cycles, Topic K4). The body responds during the cephalic phase with the release of **insulin** from the pancreas, by salivation, and the release of enzymes.

The cephalic phase is followed by the **absorptive phase**, during which nutrients are absorbed into the bloodstream, and most of the energy is stored. Cells in the liver called **glucoreceptors** detect increased blood glucose and secrete more insulin.

Amongst its metabolic effects, insulin has the following effects:

- it causes glucose to be stored in the liver and muscles in the form of a more complex molecule, **glycogen**;
- it causes the storage of lipids as **triglycerides** in **adipose tissue** (tissues found under the skin and in the abdomen, consisting of cells adapted for the storage of fats);
- it promotes the building of proteins from amino acids, and their storage in muscles. Excess amino acids are converted to triglycerides and stored in adipose tissues.

Finally, in the **fasting phase**, energy is gradually released from reserves to maintain the supply to the tissues. During the cephalic and absorptive phases, when insulin is present, all tissues are able to use glucose. However, in the absence of insulin, only neural tissues are able to absorb and metabolize glucose. When glucose levels fall the pancreas secretes a second hormone, **glucagon**, which causes the reconversion of glycogen to glucose, and its release into the bloodstream (see *Fig. 2*). Blood glucose levels are maintained within quite narrow limits by the dynamic negative feedback loops provided by these hormones. Glucagon also converts triglycerides in adipose tissues into free fatty acids and **glycerol**, from which additional glucose is formed. In the fasting phase, free fatty acids are the primary fuel for most tissues in the body, while glucose remains the main energy source for neural tissues.

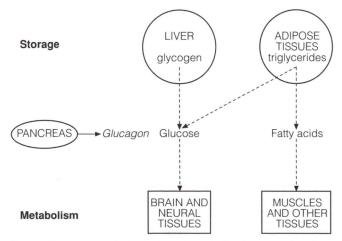

Fig 2. *Energy usage in the fasting phase. In the absence of insulin neural tissues metabolize glucose, other tissues metabolize fatty acids.*

Only about 0.5% of the energy ingested in food is stored as immediately available glycogen. Excess carbohydrates are converted into lipids in the adipose tissues. Most of this fat is kept as a longer-term store in the adipose tissues, and in an average weight person may amount to some 140 000 calories (about 85% of total stored energy). The remaining energy is stored as protein in the muscles. All of these storage processes are promoted by insulin. When glycogen levels in the liver fall, glucagon mobilizes stored fat, converting it into fatty acids and glucose. Note that the body can convert glucose and amino acids into lipids in the adipose tissues, and conversely can convert lipids into glucose. When an animal or person is starved, the glycogen and lipid reserves are used first. When these have gone proteins from the muscles are broken down into amino acids, and these are metabolized by all tissues apart from neural tissues.

M2 PHYSIOLOGICAL MECHANISMS IN EATING

Key Notes

Early theories	Cannon and Washburn's theory of hunger was that hunger and satiety are signaled by gastric contractions and their absence. But removal of the stomach or cutting the nerves from the stomach does not affect food intake. Mayer's glucostatic theory of hunger argued that the signal for eating to start is a drop of the blood glucose level below a set point, detected by glucoreceptors in a 'satiety center' in the brain. Eating ceases when it reaches that point again. Anand and Brobeck showed that bilateral lesions in a 'feeding center' in the lateral hypothalamus produce aphagia. The dual-center set-point model incorporated these two centers.
The gastrointestinal tract	Endocrine cells in the wall of the stomach and the small intestine secrete peptides, including cholecystokinin (CCK), insulin and bombesin, in response to food. These peptides have both peripheral and central actions producing satiety. The peripheral effect of CCK is mediated by autonomic afferents in the vagus nerve. Bombesin acts as a satiety signal from lower in the tract. The pancreas also secretes peptides (e.g. amylin) in response to food. These also act by stimulating afferent fibers from the gut, as well as by having direct actions on the central nervous system (CNS).
Brain mechanisms	Many aspects of feeding are organized in the brain stem. Autonomic afferents from the gut and liver enter the brain stem in the area postrema and the nucleus of the solitary tract. CCK and other peptides act here as neurotransmitters. This is the common pathway for internal and external sensory signals for hunger and satiety. These areas pass information to the lateral parabrachial nucleus in the pons, and thence to the ventromedial hypothalamus (VMH). Neuropeptide Y and other peptides are involved in the commencement of feeding, acting in the hypothalamus.

Related topics	Homeostasis (L1)	Weight control and its disorders
	Digestion, energy use and storage (M1)	(M4)

Early theories

Cannon and Washburn (1912) proposed the first notable theory of hunger. They recorded movements of the stomach by swallowing a balloon, inflating it, and connecting it to a recording device. Strong stomach contractions were accompanied by sensations of hunger: *hunger pangs*. They concluded that hunger is signaled by gastric contractions resulting from the stomach being empty, and satiety results when food reaches the stomach, stopping these hunger contractions. It was soon shown that this is not an adequate explanation. First, surgical removal of the stomach in humans does not prevent feelings of hunger, and patients maintain normal body weight (Wangensteen and Carlson, 1931).

Second, cutting the nerves between the stomach and the central nervous system does not affect food intake (Morgan and Morgan, 1943).

Mayer (1953) proposed a **glucostatic theory** of hunger, arguing that the signal for eating to start is a drop of the blood glucose level below a set point (see Topic L1). Eating ceases when it reaches that point again. In each case, the sensors for blood glucose levels were proposed to be **glucoreceptors** in the brain. Destruction of these glucoreceptors in rats by the injection of *gold thio-glucose*, which binds to the receptors and then kills them, produced **hyperphagia** (overeating); the rats ate until they become extremely obese. Subsequent histology of the rats' brains revealed that the destroyed cells were in the **ventro-medial hypothalamus (VMH)**. Mayer labeled this area a **satiety center** for eating. People with tumors in this region sometimes develop hyperphagia, which supports this view.

Anand and Brobeck (1951) showed that bilateral lesions in the **lateral hypo-thalamus (LH)** produced **aphagia** (failure to eat), and they concluded that this was a feeding center that controls the start of eating. Subsequent studies showed that electrical stimulation of the LH could produce eating. Following this in the 1950s and 1960s, the predominant view of the control of eating was the **dual-center set-point model** (see *Fig. 1*). In this model eating commences with the stimulation of the LH feeding center by decreased blood glucose, and ceases with the stimulation of VMH satiety center by blood glucose above the set point. The satiety center acts by inhibiting the feeding center.

For several reasons this model has to be rejected. Closer examination of the behavior of animals after VMH lesions reveals that the result is not simply endless hyperphagia (see *Fig. 2*). The huge increase in food intake is followed after about 12 days by recovery to a steady level. Weight is then maintained at a higher level, although food intake is only slightly higher than normal. Starvation at this stage leads to a fall in weight, but the higher level is reached again when the animal is given free access to food. Conversely, following force

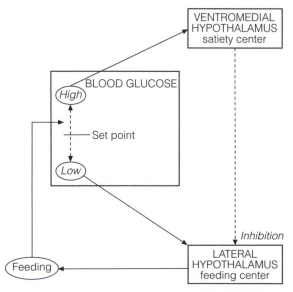

Fig. 1. The dual-center set-point theory. Low blood glucose stimulates the LH to start feeding. High blood glucose stimulates the VMH to inhibit the LH, ending feeding.

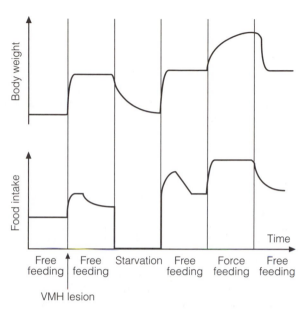

Fig. 2.   Food intake and body weight in rats following VMH lesions. Under free feeding the animal maintains a higher body weight.

feeding, the animals reduce their food intake, again until their weight is restabilized (Teitelbaum and Stellar, 1954). So, VMH lesions do not interfere with the control of food intake. Rather, they seem to raise the set point for body weight. The original studies probably damaged a nerve tract that mediates autonomic control over the pancreas, increasing insulin production and decreasing glucagon secretion. This causes more energy to be stored in the form of body fat, and makes it less accessible.

Furthermore, LH lesions do more than produce changes in eating, they also affect drinking and other behaviors. Most LH-lesioned rats eventually start to eat and drink again (as long as they are kept alive long enough by being given food and water through a tube). Just as VMH-lesioned rats stabilize their weight at a higher level, so LH-lesioned rats maintain their weight at a lower level.

The gastrointestinal tract

It has long been known that the stomach can provide a satiety signal, since injection of food directly into the stomach of hungry experimental animals stops them eating. Conversely **sham feeding**, in which animals are given fistulas preventing food from reaching their stomachs, results in copious eating. This sham feeding is stopped by injections of food directly into the stomach, while removal of food from the stomach starts the animal eating again (Davis and Campbell, 1973). The original interpretation of this was either that nutrients absorbed into the blood from the intestines trigger satiety, or that there is some direct sensory signal (perhaps stretch receptors in the stomach) that has this effect. However, several facts indicate that these cannot be the only mechanisms:

- preventing the passage of food from the stomach does not arrest the satiety mechanism (Deutsch and Gonzalez, 1980);
- pressure in the stomach does not increase following eating (Young and Deutsch, 1980);

- denervation of the stomach does not prevent the cessation of feeding (Gonzalez and Deutsch, 1981).

Koopmans (1981) transplanted a stomach into a rat, and connected it to the rat's blood supply. Food injected into the transplanted stomach decreased eating. Since the stomach was not connected to the rat's nervous system, and since nutrients are not absorbed from the stomach, the satiety signal must come from a chemical signal produced by the stomach itself.

Endocrine cells in the wall of the stomach and the small intestine are known to secrete **peptides** (chains of amino acids smaller than proteins) in response to food. These peptides include **cholecystokinin (CCK)**, **insulin** and **bombesin**, although there are many others. These peptides have both peripheral and central actions producing satiety. Peripheral injection of CCK into rats and humans causes them to stop eating (Smith and Gibbs, 1994). This peripheral effect of CCK is mediated by autonomic afferents in the **vagus nerve** (Guan, Phillips and Green, 1996). The same is true of bombesin (Yoshidayoneda et al., 1994). The pancreas secretes peptides (e.g. **amylin**), in addition to insulin, in response to food. These also act by stimulating afferent fibers from the gut, as well as by having direct actions on the CNS. Insulin crosses the blood–brain barrier (see Topic C1), and attaches to receptors in the hypothalamus. However, in the levels actually found in the body, insulin has no short-term effect on hunger or food intake in humans (Chapman et al., 1998). Its long-term effects are likely to be secondary to its metabolic effect on glucose.

Brain mechanisms

Many aspects of feeding are organized at the level of the brain stem. Decerebrate rats (with the brain stem separated from the forebrain) can eat. They can make food selections based on taste, and can respond to peripheral signals of hunger and satiety outlined above. The brain stem contains receptors for insulin and other neuropeptides involved in the control of feeding. In these animals, neural mechanisms in the brain stem are disconnected from the circuits in the cerebral hemispheres that normally control them (see Grill and Kaplan, 2002).

Autonomic afferents from the gut and liver enter the brain stem in, amongst other places, the **area postrema** and the **nucleus of the solitary tract** (NST). High concentrations of CCK and other peptides are found in these locations, where they act as neurotransmitters. Selectively blocking central CCK receptors has a much greater effect on food intake than blocking peripheral CCK receptors alone. Lesions in these areas prevent the feeding that usually takes place in response to sudden prevention of mobilization of glucose and lipids. Since these regions also receive sensory afferents from the gustatory and olfactory systems (see Topic I2), it appears that this is the common pathway for internal and external sensory signals for hunger and satiety. These areas pass information to the **lateral parabrachial nucleus (LPN)** in the pons, and thence to the VMH. It appears that hypothalamic and brain stem mechanisms are together involved in the control of energy balance.

Peptides are involved in the *commencement* of feeding. **Neuropeptide Y (NPY)** is found in many parts of the brain, especially in the **paraventricular nucleus (PVN)** and other parts of the hypothalamus, and in brain stem areas such as the NST and the LPN. Injection of NPY into the hypothalamus causes sated rats to start feeding (Clark et al., 1984). It has been shown that rats that become obese after being given free access to a varied and highly palatable diet have increased NPY in the PVN at an early stage in their weight gain. However, this subsequently returns to normal although weight gain continues. Rats that did not become obese under the same regime showed normal NPY, but increased amounts of another peptide, **galanin**, in the **arcuate nucleus** (Pedrazzi et al., 1998). The arcuate nucleus sends fibers to the PVN, probably using galanin as neurotransmitter. An alternative view is that NPY mediates hunger for carbohydrates, while galanin is involved in hunger for fats. Injection of a serotonin antagonist increases the amount of NPY in the hypothalamus, indicating that the serotonergic system acts by inhibiting the NPY mechanism. The feeding-inducing effects of NPY are reduced by the injection of the hormone **leptin**, (see Topic M4) into the cerebral ventricles, where it acts on the PVN (Kotz et al., 1998).

More and more peptides and proteins are being found to influence feeding. For example **ghrelin**, produced in the stomach and gut, and **orexin**, produced amongst other places by neural axons in the hypothalamus, interact with NPY to stimulate feeding (Toshinai et al., 2003).

The complexity of this is indicated further by experiments on a strain of rats, **Zucker rats**. These carry a gene that causes hyperphagia and obesity. Cole, Berman and Bodner (1997) have pointed out that mechanisms based on opioids, norepinephrine, dopamine, NPY, 5-HT, histamine, galanin and CCK have all been shown to be involved in the production of obesity in these animals. Much remains to be established about the physiological control of feeding. However, it can be seen to be a complex interaction of neurotransmitters, hormones, peripheral chemical and sensory factors, and brain nuclei.

M3 DIETARY CHOICE AND PSYCHOLOGICAL FACTORS

Key Notes

Palatability	We eat more when the food tastes nice: positive incentive value. When deprived the palatability of food increases. The rewarding properties of eating may be due to the release of endogenous opioids in the hypothalamus. They also involve dopaminergic systems in the median forebrain bundle and neurons in the orbitofrontal cortex.
Sensory-specific satiety	Sensory-specific satiety is satiety to a particular food. It is specific to the flavors of foods, and is based on the sensory properties of foods. It facilitates variety in diet. Sensory-specific satiety is an example of habituation. Neurons in the lateral hypothalamus are responsive to the incentive properties of food, and reflect sensory-specific satiety.
Food preferences	Neonates show preferences when given sweet liquids, and aversive responses to sour and bitter tastes. The learning of further taste preferences starts soon after birth. Flavor–flavor learning results from the association of a new flavor with one that is already preferred. In flavor–nutrient learning a new preference is established by pairing a food with another food that has nutritive consequences. Much of our learning of food preferences takes place within the family. Less palatable foods are often associated with negative affect, while palatable foods are presented in pleasant contexts.
Learning what not to eat	We also learn what not to eat, through conditioned aversion. This results from pairing a food with internal ill effects, such as nausea. Conditioned aversion shares some of the properties of classical conditioning, but has important differences. These include one-trial learning, learning after long delays, and not occurring with external punishment such as electric shocks. Conditioned aversion may be based on simple neural circuits in the nucleus of the solitary tract and the parabrachial nucleus, but also involves the cerebral cortex.
Learning when to eat	We learn to eat at particular times of day, and when stimulated by sounds and sights associated with food.
Related topic	Weight control and its disorders (M4)

Palatability   The results we saw in animals with VMH lesions (see Topic M2) depend on the animals being provided with a diet that is not only nutritious, but also palatable. If the palatability of the food available to the animals is low, then the hyperphagia is less, and the new stable weight is not much higher than that of unoperated animals. Generally, we eat more readily, and also eat more, when the food tastes nice. Palatable food is said to have **positive incentive value**, and nicer tasting food has a higher incentive value. Under conditions of deprivation, the palatability of any food increases. Opioid antagonists reduce pleasantness ratings of foods, suggesting that the rewarding properties of eating are due to the release of **endogenous opioids** in the hypothalamus (Le Magnen, 1990).

The neural mechanisms of palatability have been investigated using *taste reactivity* studies. For example, Pecina and Berridge (1996) showed that feeding and positive hedonic responses induced by microinjections of benzodiazepine are greatest when the injections are into the fourth ventricle, in the brain stem. This suggests that brain stem mechanisms based on **benzodiazepine–GABA** receptors are responsible for some effects of the palatability of food. Later studies have shown this mechanism to be located in the parabrachial nucleus of the brain stem (Soderpalm and Berridge, 2000).

Palatability is related to the rewarding properties of food; that is, food given as a reward can promote learning. The general dopaminergic neural mechanisms for this are outlined in Topic P2. In addition, neurons in the **orbitofrontal cortex** respond differently to different specific rewards. This seems to work with the dopaminergic system to shape learning.

Sensory-specific satiety

Eating a particular food has a number of consequences.

- during consumption, the rate of eating slows;
- this is followed for up to an hour by a decrease in palatability; and
- it becomes less likely that the same food will be chosen again.

These effects are all specific to the particular food eaten, and are known as **sensory-specific satiety** (see Hetherington and Rolls, 1996). It can be observed during a meal: even if you feel completely sated by the main course, you might 'find room' for dessert, even though it contains more calories than you have already eaten.

Sensory-specific satiety occurs in sham-eating animals (see Topic M2; Swithers and Hall, 1994). It does not, therefore, depend on nutrients or other chemical signals from the digestive system. It is specific to the flavors of foods, and must, therefore, be based on the sensory properties of foods. The only macronutrients that are detectable by taste are the sugars; the apparently distinctive flavors of fats and proteins come from molecules associated with them in foods. Of the other nutrients, only salt has a specific gustatory (taste) sensation. Nevertheless, we presume that the adaptive function of sensory-specific satiety is to facilitate variety in diet. Since flavor and nutrient composition are generally correlated in foods, sensory-specific satiety helps to ensure variety of nutritional intake. Hetherington, Pirie and Nabb (2002) showed that sensory-specific satiety is essentially a short-term mechanism of food-intake control. Pleasantness and desire to eat a food declined over days of repeatedly eating the same food (even chocolate), but the amount eaten did not change.

Sensory-specific satiety is an example of **habituation**: a type of non-associative learning (Topic P1). Rolls (1993) showed that cells in the lateral hypothalamus originally thought to be responsive to food (see Topic M2) are actually responsive to the *incentive* properties of food, and reflect the process of sensory-specific satiety. Although their activity declines in the presence of food that has just been eaten, they remain active to different food, to the sight of different food, and to stimuli conditioned to food.

Food preferences   Newborn infants show distinctive facial responses to sweet, sour, bitter, and possibly salt substances in the mouth. The responses to sour and bitter tastes involve expulsive mouth movements. Newborns also show taste preferences, shown by faster sucking rates and/or greater food intake when given sweet liquids. Most studies of the gustatory responses of newborns show no specific response to salt, and in preference tests infants seem to be indifferent to salty tastes. There is evidence for aversive responses to sour liquids, and strongly bitter tastes (Mennella and Beauchamp, 1996). These might have evolved to protect us against eating injurious substances; many poisons have a bitter taste. Similarly, the pleasantness of sweet tastes might indicate foods that are good to eat.

Beyond these innate responses, animals and people learn which foods are good to eat, and which are not. One example of this is the development of preference for salty tastes during the first six months of life (Harris and Booth, 1987). There is evidence for prenatal influences on food preference in other species, but in humans the learning of taste preferences starts soon after birth. Early sodium deficits lead to a greater salt appetite when children are older (Beauchamp and Cowart, 1993). Flavors of foods eaten by a mother are transferred to her milk, and this can influence the offspring's food preferences (Mennella and Beauchamp, 1996).

Associative learning (Topic P1) provides the most powerful source of food preferences. **Flavor–flavor learning** results from the association of a new flavor with one that is already preferred. The most obvious example of this is the association of flavors with sweetness. In one study people expressed a preference for unfamiliar varieties of tea that were sweetened over those that were not. This preference continued even when they were subsequently tried unsweetened, demonstrating that they had learned the preference by the association of the flavor of the tea with sweetness (Zellner et al., 1983).

In **flavor–nutrient learning**, a new preference is established by pairing one food with another food that has nutritive consequences. In one study (Sclafani and Nissenbaum, 1988) rats were given water with one of two different flavors, and it was arranged that each time they took a drink from one particular flavor, a nutritious solution was delivered directly to the stomach. After four days, the animals were given a choice between the two flavors, and chose the one that had previously been associated with the nutritious result. The animals had learned which taste indicated the supply of nutritious food.

One further, very important, source of learning in diet is direct **parental influence**. The foods that we find most palatable without learning are sweet foods, and 'healthy' foods are less palatable. The adoption of a healthy diet, therefore, must be based on learning, and much of this learning takes place within the family. Within the family, the eating of less palatable foods is often associated with negative affect (coercion, threats or actual punishment), while inherently palatable foods, high in carbohydrates often combined with fats, are presented in pleasant contexts (parties, treats, etc.). This makes it difficult for the child to learn healthy eating habits, particularly as it is combined with **neophobia**: a distrust of trying new foods that is common in children and in other species, presumably as a protective device.

Learning what   We also learn what *not* to eat, through the process of **conditioned aversion** (or
not to eat   **taste aversion learning**). In an early study (Garcia and Koelling, 1966) rats were given saccharine to taste, and were then injected with a lithium salt, which

induces distress (lithium induces nausea in humans). After only one such experience, the rats learned to avoid saccharine solutions. Aversion can be conditioned to tastes, or to the smell, sight or thought of particular foods. Many people develop an aversion to foods that have made them ill. The significance of conditioned aversion is that a food that has once made an animal ill is likely to do so again, and so should be avoided.

Conditioned aversion is not simple classical conditioning (Topic P2), although it shares some of the same properties:

- the aversive response shows *generalization* to similar tastes;
- the response declines the more different the tastes are from the conditioned one (a *generalization gradient*);
- the aversive response shows *extinction* after repeated, non-reinforced presentations of the original food.

However, conditioned aversion shows important differences from classical conditioning:

- the response is formed in a single trial – in classical conditioning a response becomes associated with the conditioned stimulus only gradually;
- it can take place when the food and the sickness are separated by a long interval of up to 12 hours; in classical conditioning, delays of only a few seconds usually prevent the formation of conditioned responses;
- it does not occur with external punishment such as electric shocks.

These differences are crucial if conditioned aversion is to have a protective effect. An animal could not learn to avoid poisonous food over a number of pairings of food and sickness; the learning *must* take place in one trial. Similarly, the ill effects of poisonous food are frequently delayed by hours.

Garcia, Hankins and Rusiniak (1974) suggested that conditioned aversion may be based on simple neural circuits in the **nucleus of the solitary tract** and the **parabrachial nucleus**, which are the sites of convergence of visceral and taste afferents. However, it is clear that other locations in the central nervous system are involved, including the **lateral hypothalamus** (Caulliez, Meile and Nicolaidis, 1996) and the **insular cortex** in the temporal lobe (Bermudezrattoni and McGaugh, 1991).

Learning when to eat

We also learn *when* to eat. We tend to eat at particular times of day. Rats can quickly learn to associate regular times of day with feeding, and they secrete insulin in anticipation at those times of day (Woods, 1995). Rats that have been conditioned to associate the availability of food with an audible stimulus will eat more when that stimulus is repeated, even when food is continuously available between stimuli (Weingarten, 1983). People isolated from the normal cues to the time of day (see Topic K4) often change their eating patterns to fewer, larger meals. Thus, eating (or hunger) and the glandular components of the cephalic phase of digestion are triggered by internal circadian rhythms, by external stimuli associated with daily activity cycles, and by specific stimuli like the sound of crockery and cutlery.

M4 WEIGHT CONTROL AND ITS DISORDERS

Key Notes

The regulation of body weight

Early attempts to explain the regulation of body weight proposed that it is controlled by a set-point mechanism based on body fat. It is better to view weight control as depending on a variable settling point: a level at which the multiple factors that influence body weight reach equilibrium. Force feeding and dieting lead to new higher and lower settling points, respectively. The new settling points are maintained, respectively, by decreased or increased metabolic efficiency. Part of the change in efficiency derives from changes in activity in brown adipose tissue.

Obesity

Obesity means having more than average body fat and ultimately results from taking in more calories than we use. Factors causing this include genetic susceptibility, eating more calories than required, the availability of highly palatable foods, the division of meals into more than one course, and the use of energy-dense foods as rewards. Whatever method is used, weight loss is rarely maintained long term. Obesity is also maintained by the yo-yo effect, which results from the increased metabolic efficiency that occurs after severe calorie restriction.

Anorexia nervosa and bulimia nervosa

The central features of anorexia nervosa (AN) are refusal to maintain body weight within the normal range, fear of gaining weight, low self-esteem, overestimation of body size, and loss of menstrual periods. The extreme weight loss can be fatal. Weight loss is mainly achieved by self-starvation (restricting type anorexia nervosa). Up to half of sufferers engage in binge eating and/or purging (binge-eating/purging type anorexia nervosa). The main characteristic of bulimia nervosa (BN) is binge eating, accompanied by compensatory vomiting or laxative use. Factors involved in the etiology of eating disorders include sociocultural factors (including media representations of the 'ideal' female figure), family pressure and interaction patterns, perfectionism, and personal and physiological factors. There are genetic predisposing factors in both AN and BN. Many studies have reported particular physiological changes or abnormalities in anorexic persons. But these might result from, rather than cause eating disorder.

Related topics

Physiological mechanisms in eating (M2)

Dietary choice and psychological factors (M3)

The regulation of body weight

The mechanisms that we examined in Topic M2 concern the control of *feeding*; that is they control the eating of meals. But the long-term result of ventromedial hypothalamus lesions is a fairly small change in food intake, but a large change in body weight. Perhaps we have mechanisms for controlling body weight, in addition to those controlling the start and end of each meal? Initial attempts to explain this proposed that meal size is controlled by a set-point mechanism based on glucose levels, while body weight is controlled by a set-point mechanism based on body fat. In this view, the amount of fat stored in the body has a fixed target level. If fat levels drop below this set point (and body weight drops) feeding will take place to increase fat deposits to the set point. Conversely, if the fat content of the body rises above the set point, the animal will eat less to reduce fat deposits.

It is obvious, though, that human body weight varies enormously, both between and within individuals. This rules out a rigid, homeostatic model, with body weight maintained at a fixed set point. It is better to view weight control as depending on a variable **settling point**. The settling point represents a level at which the multiple factors that influence body weight reach an equilibrium. The main distinctions between set-point and settling-point models are:

- set points are fixed, but settling points are variable;
- mechanisms that control set points return the system to the set value, but the mechanisms that control settling points simply limit the extent of change.

One example of the self-limiting nature of the settling-point mechanism is shown in the effect on body weight of feeding a low or a high calorie diet (see *Fig. 1*). An animal (or human) fed a low calorie diet will lose weight. However, the rate of weight loss gradually declines until after several weeks body weight is maintained at a new, lower level. Conversely, a high calorie diet produces a diminishing increase in body weight until a high settling point is reached, when weight is maintained at that level. (Compare the effects of ventromedial hypothalamus lesions in Topic M2.)

A similar effect results from making available a highly palatable, high calorie diet. In one study (Sims and Horton, 1968) prisoners were given such meals several times each day. All increased their calorie intake enormously. Despite

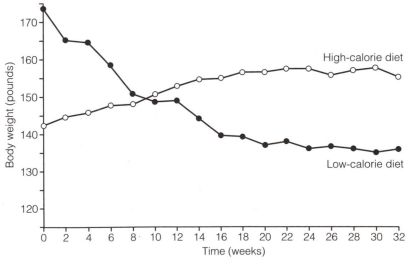

Fig. 1.   Settling points. People following high and low calorie diets reach new steady body weights, despite maintaining the high and low calorie intakes.

this, they only gained a small amount of weight, and at the end of the experiment they returned to their usual diets and their previous weights. The main process that maintains settling points is variation in **metabolic efficiency**. In all the examples we have looked at where body weight stabilizes under increased calorie intake, more energy is consumed (see Martin, White and Hulsey, 1991). This is shown by increased heat production. Conversely, when loss of weight ceases during a low calorie diet, the efficiency of energy use is improved (Rothwell and Stock, 1982). At least part of the variation in metabolic activity derives from changes in activity in **brown adipose tissue**. This tissue is specialized to generate heat.

Obesity

The term **obesity** may be used simply to refer to having more than average body fat. A body weight more than 20% above average is defined as *medically significant obesity*, contributing to a wide range of illnesses, especially cardiovascular disease (Rand, 1994). Obesity results only, in the last analysis, from taking in more calories than we use. However, there are many possible reasons for such an imbalance. One obvious factor is the wide availability of processed foods of high palatability in western societies. This can increase calorie consumption and may change the settling point so that body weight stabilizes at a higher level. But not everybody exposed to such a diet becomes obese.

One undoubted factor in obesity is learning. In many families children are encouraged, or forced, to eat all the food placed in front of them. This can lead to eating more calories than required. If it becomes habitual it can lead to obesity. The division of meals into more than one course exacerbates this, allowing the release from sensory-specific satiety to increase total calorie intake (see Topic M3). *Energy-dense foods*, foods with a high concentration of calories per unit weight, usually fat-based foods, are highly rewarding, particularly if sweet. The association of sweetness with high fat content of many processed foods is the basis for learning a preference for these foods. Studies using opiate receptor blockers (Drewnowski et al., 1995) have shown that this preference is based on endogenous **opioid** circuits (presumably β-**endorphins**; see Topic F3). This can account for phenomena related to 'comfort-eating', when food is used to elevate mood.

There are also genetic factors. The weight of people who have been adopted in infancy is more highly correlated with the weight of their natural parents than that of their adoptive parents (Stunkard et al., 1986). A number of possible sites on human chromosomes have been identified as contributors to the likelihood of a person becoming obese (Comuzzie, 2002). Strains of mice have been bred with a genetic predisposition to obesity. These laboratory mice have mutations in genes that encode either for **leptin** or for a receptor for leptin. Leptin is a hormone secreted by fat cells. Amongst other effects it attaches to receptors in the hypothalamus where it counteracts **neuropeptide Y** (see Topic M2), thereby suppressing appetite and increasing energy expenditure. Levels of leptin in humans are correlated with amount of body fat (Considine et al., 1996). Leptin appears to be crucially involved in the control of body fat levels through its effects on appetite and energy expenditure. Disorders of leptin production, or of sensitivity to circulating leptin, might be a causal factor in human obesity.

It is partly because of the hereditary contribution that obesity is difficult to treat. Whatever method is used, ranging from calorie-reduced diets to surgical removal of fat and reduction of the size of the stomach, weight loss is rarely maintained in the long term. Obesity is also maintained by the **yo-yo effect**.

This follows the increased metabolic efficiency that occurs after severe calorie restriction. It may occur in people who lose weight by calorie restriction, making it harder for them to avoid putting weight on again. In one study, Brownell et al. (1986) fed rats a high calorie, palatable diet until they became obese. On average, it took 46 days for the rats to reach the criterion of obesity. The animals were next starved until their weights returned to normal, and then again fed on the high calorie diet. This time, it took only 14 days for them to become obese. The implication is that their metabolic efficiency had increased, so that more energy was stored as fat.

Anorexia nervosa and bulimia nervosa

The central features of **anorexia nervosa (AN)** are refusal to maintain body weight within the normal range, fear of gaining weight, low self-esteem coupled with distorted perception (overestimation) of body size, and loss of menstrual periods. The extreme weight loss in AN leads to a range of physical problems and can be fatal. Despite the name, which literally means *loss of appetite of nervous origin*, sufferers are often preoccupied with food. It is much more common in females, about 1% of whom will have the illness at some time. It can start in childhood, but most commonly between 14 and 18 years of age. Weight loss is mainly achieved by self-starvation, and when confined to that the illness is described as **restricting type AN**. Up to half of sufferers engage in binge eating and/or purging (self-induced vomiting, or use of laxatives or enemas), and may be diagnosed as **binge-eating/purging type AN**.

The main characteristic of **bulimia nervosa (BN)** is binge eating, accompanied by compensatory calorie-losing behavior which may allow the maintenance of a normal body weight. The compensatory behaviors are most commonly vomiting or laxative use, but may include fasting and strenuous exercise. The vast majority of those with BN are women, with onset usually between 15 and 19 years of age. Up to 6% of young women will suffer from BN, although more will occasionally purge, and up to 50% admit to one or more episodes of binge eating. The binges in BN usually start with feelings of tension or anxiety, which are relieved by the consumption of huge quantities of food. But the relief of tension is followed by guilt and depression, followed by purging, apparently as a way of relieving these feelings, and of avoiding weight gain. However, neither vomiting nor laxative and enema use prevent the absorption of calories. The purging temporarily relieves the guilt and depression, rewarding the whole binge–purge cycle. However, the person with bulimia nervosa develops low self-esteem, particularly feeling powerless and disgusted with themselves.

There have been many theories for the etiology of eating disorders (see Marx, 1994). Currently, multifactorial views are in favor, including sociocultural, family, personal and physiological factors. The non-physiological factors include the thinness of the 'ideal' female figure as depicted in the media; pressure from the families of eating disorder patients towards academic and personal achievement; the development of perfectionism in the child; and disturbed patterns of interaction in the families of eating disorder patients, particularly concerning the expression of emotion. Some see the development of eating disorder as a way in which the young person can gain control over herself.

Family incidence studies have suggested genetic predisposing factors in both AN (Ben-Dor et al., 2002) and BN (Bulik et al., 2003). AN is about 12 times more frequent in people with a close family member who has AN, and BN about 4 times more frequent. No certain chromosomal basis is known for these familial occurrences, although, as with most behavioral states, multiple influences are likely. Bulik et al. (2003) have found a link with a site on one particular chromosome, 10p. Vink et al. (2001) found that 11% of AN sufferers share a mutation in a gene that codes for a protein, AGRP, known to stimulate eating in rodents.

N1 REPRODUCTION AND SEXUAL DIFFERENTIATION

Key Notes

Reproduction	Sexual reproduction is the usual way of reproducing in vertebrate species. It passes only half of the genes from each parent to the offspring. The mixing of genes confers greater resistance to environmental factors, in particular to parasites such as viruses.
Mating patterns	Mating patterns and the investment each parent has in offspring determine the reproductive role played by each parent. In mammals, the female has a much greater investment in each of the offspring than the male. Sociobiologists argue that much modern human behavior has its origins in such gender differences. Sexual behavior varies enormously between different species. In many species, including rodents, sexual behavior is highly stereotyped, and the behavior of males and females is distinctive. Human sexual behavior is much less stereotypical, and less controlled by hormones.
Sexual differentiation	Males and females show sexual dimorphism. This comprises primary sex characteristics apparent at birth, and secondary sex characteristics that develop later. Males and females are differentiated by females possessing two X chromosomes, and males one X and one Y chromosome. In humans, six weeks after fertilization the Y chromosome produces HY antigen. This causes male embryos to develop testes, which produce testosterone, and anti-Müllerian hormone (AMH). Testosterone causes the development of the male genitalia and AMH prevents the development of the female genitalia. In females, the absence of these substances allows the development of female genitalia. Changes at puberty are mostly produced by androgens in males, and estrogens in females.
Related topics	Psychology and biology (A1) Parental behavior (N4)

Reproduction   Many biologists argue that the prime underlying motivation of all organisms is to maximize the proportion of their own genes in the next generation. The most straightforward way to do this would be to reproduce alone, thereby passing all of your genes to your offspring. However, such **asexual reproduction** is extremely rare in vertebrate species. In **sexual reproduction** only half of the genes from each parent are passed to the offspring. For such a mechanism to have evolved there must be an enormous advantage in each generation being different from the previous one. Mixing genes from two parents produces an enormous variety of genotypes (see Topic A1). Different combinations of genes provide different responses to environmental factors, in particular to parasites such as viruses. Viruses need to 'recognize' a particular protein in the cell wall in order to infect a cell. If the proteins change in successive generations the virus will be less able to infect the cell and weaken or kill the host. This increases the chances that some offspring (and hence copies of half of an individual's genes) will survive (see Ridley, 1994).

Mating patterns   Sexual reproduction involves sexual behavior. Vertebrate species show a number of different mating patterns. Amongst mammals the most common reproductive pattern is known as **promiscuity**. The characteristics of this reproductive system are:

- females are highly selective in their choice of mate;
- most males never mate;
- most females mate with several males.

Although human mating systems show great variability, most are described as **monogamous**, denoting that a male and female form an exclusive pair. Monogamy is relatively rare in mammalian species. As with other species, humans may show **serial monogamy**, meaning that partners may be changed. However, in humans, as well as in monogamous species like many birds, infidelity is common. That is, while staying as a pair the male, the female, or both will copulate with other partners outside the pair (see Baker and Bellis, 1995). Other, rarer, mating systems are **polygyny**, in which one male mates with a 'harem' of females, and **polyandry**, in which each female mates with several males, but each male mates with only one female.

The mating pattern helps to determine the role played by each parent in the reproductive process. Another factor is the biological investment each parent has in the offspring. In mammals, the female carries offspring internally for an appreciable time, and cannot conceive again until after these offspring are born. Furthermore, she gives birth to infants that are totally dependent on her for sustenance. Therefore, the female has a much greater investment in each of the offspring. In mammals, and many other animals, females produce relatively few **gametes** (reproductive cells; for females these are ova). Males, on the other hand, produce many millions of gametes (sperms), each of which costs far less in resources to produce than each ovum. Furthermore, males do not have to wait before they breed again, and they are not in many species necessarily involved in the care of the offspring. Sociobiologists argue that much modern human behavior has its origins in such gender differences in parental investment of time and physical resources.

Sexual behavior varies enormously between different species. For example, some animals have an annual breeding season, usually with both sexes pursuing mating only at one time of the year. Others, including rodents, have a more frequent cycle, with males interested in mating throughout the year, but with females receptive only for some of the time (the **estrus cycle**). In others, including humans, both sexes are sexually active throughout the year. In many species, including the rodents commonly studied in the laboratory, sexual behavior is highly stereotyped, and the behavior of males and females is distinctive. Rodent males, and at the appropriate time females, show attraction to animals of the opposite sex. The females show species-typical **proceptive behaviors**, which include running towards and away from the male in a particular way, known as *hopping and darting*. She only does this when **progesterone** levels are high, at the appropriate phase of the estrus cycle. Staying near to the female, usually sniffing her, shows the male's attraction. He will try to mount the female, and in response to tactile stimulation on her flanks, the female's proceptive behaviors are replaced by **receptive behavior**, characterized by a particular stance called **lordosis**, which permits copulation to occur. Human sexual behavior is much less stereotypical than that of the animals we study most.

Sexual differentiation

Men and women usually look different, as is the case with males and females of many other species. Some differences, the **primary sex characteristics**, are apparent at birth (e.g. the external genital organs), some are there not externally visible (e.g. the internal genitalia), and other differences appear at puberty (the **secondary sex characteristics**). There are also behavioral differences. Together, all of these differences are referred to as **sexual dimorphism**.

In most species, males and females are genetically differentiated by the fact that females possess two X chromosomes, while males possess one X and one Y chromosome. For about six weeks after fertilization, male and female human embryos are otherwise identical. Each contains undifferentiated (primordial) **gonads**, with different tissues in an inner medulla and an outer cortex. Each also has two sets of ducts; the **Wolffian system** and the **Müllerian system** (see *Fig. 1*). Human sexual dimorphism stems almost entirely from a single gene on

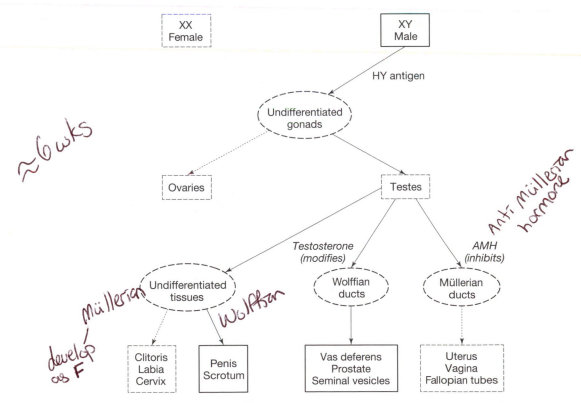

Fig. 1. Human sexual differentiation. In females, the tissues indicated with oval borders develop into the organs shown with broken lines. In males, the HY antigen produced in response to a gene on the Y chromosome initiates the sequence of changes indicated, producing the organs shown with full lines.

the Y chromosome. At six weeks this produces **HY antigen**. This causes the medulla of the undifferentiated gonads of male embryos to develop into testes (see *Fig. 1*). The testes then produce two substances, the androgen hormone **testosterone**, and **anti-Müllerian hormone** (**AMH**; also known as Müllerian inhibiting substance). Testosterone causes the medulla of the primordial gonads and the Wolffian system in the male embryo to develop into the male genitalia (seminal vesicles, vas deferens, scrotum, penis, etc.). Simultaneously AMH prevents the cortex and the Müllerian system developing into the female internal genitalia in males. In the female, the *absence* of these substances allows the development of female genitalia (uterus, fallopian tubes, labia, clitoris, etc.). We will see later (Topic N3) that hormonal differences whilst in the uterus also produce differences in the central nervous system.

Sexual differentiation in the embryo depends almost entirely on the presence or absence of testosterone. However, changes at puberty are mostly produced by hormones whose action is specific to the sex of the individual: *androgens* in males, and *estrogens* in females. Testosterone leads to the secondary sex characteristics of the male adult body (beard growth, increased muscle bulk, growth of external genitalia, enlargement of vocal cords and larynx), while estrogens from the ovaries produce the female changes (breast development, subcutaneous fat deposits, onset of menstruation). The exceptions to this are that the growth of pubic and axillary hair in both sexes results from androgens (which, in females, are secreted mainly by the adrenal cortices; see Topic E6).

Such effects of hormones, producing changes preparing the body for particular activities, are known as **organizing** (or organizational) **effects**. However, not all organizing effects are so obviously structural, and we will look at behavioral organizing effects in Topic N2.

N2 HORMONAL CONTROL OF SEXUAL BEHAVIOR

Key Notes

Activating effects of hormones in males	Stereotyped sexual behaviors only occur in the presence of 'sex-appropriate' hormones. This is an activating effect of hormones. Individual differences in sexual activity are not related to adult testosterone levels. Sexual behavior of a male rodent depends on the presence of testosterone and of external factors. The presence of a receptive female animal increases the male's testosterone. Castration of adult men is more variable than that of rats. Some men become asexual within weeks or months while others retain full sexual capacity and interest.
Activating effects in females	The readiness of females of many species to engage in reproductive behavior depends on cyclic release of gonadal hormones. Ovariectomy in female rodents produces rapid loss of sexual behavior. The reinstatement of the behavior requires matching the sequence of hormone production. Neither ovariectomy nor the loss of female hormones after the menopause directly affects sexual behavior in human females. The sexual drive of women is more closely affected by androgens (secreted by the adrenal cortex and the ovaries).
Organizing effects of testosterone	The development of both male and female sexual behavior depends on the perinatal presence or absence of androgens. Testosterone has organizing effects, preparing infant animals for male sexual behavior, and reducing the occurrence of female sexual behavior: masculinization and defeminization. In the absence of testosterone, animals develop female patterns of behavior and not male ones: feminization and demasculinization. Evidence from pathological conditions suggests some similar effects in humans.
Related topics	Neural mechanisms in sexual behavior (N3) Aggression (O4)

Activating effects of hormones in males

The stereotyped sexual behaviors described in Topic N1 only occur in the presence of 'sex-appropriate' hormones (androgens in males, and estrogens in females). Removing the gonads of adult animals completely abolishes the behaviors within days, and the injection of sex-appropriate hormones reinstates them immediately. Injecting sex-*inappropriate* hormones has little or no effect. In animals that have annual breeding cycles, testosterone production almost stops during parts of the year when breeding does not take place. This type of hormone effect is called an **activating effect**, since the hormone results in the activation of programmed behaviors.

Male rodents show wide individual differences in sexual activity. These differences are not related to adult testosterone levels. After castration, if sexual activity is reinstated by testosterone injection, the amount of sexual activity does not depend on the amount of hormone injected. Instead, the original individual differences occur (Grunt and Young, 1953). Furthermore, previous levels of sexual activity can be reinstated by as little as one-tenth of the amount of testosterone before castration. Giving more testosterone does not increase sexual activity. Testosterone, then, shows a *threshold effect*: once the animal has a certain, critical level, it can, and will, engage in sexual activity, given the appropriate external stimuli. This is also described as a *permissive* effect of a hormone. A certain level of a hormone permits a particular behavior to take place, but the actual occurrence of that behavior depends on other factors.

Sexual behavior of a male rodent depends on the presence of external factors (a proceptive and receptive female conspecific), and of internal factors (the perinatal and developmental organizing effects of testosterone, together with the current activating effects of testosterone). The presence of a female animal in the receptive stage of the estrus cycle produces a measurable increase in the male's testosterone level. These external and internal factors are interrelated. The male animal's behavior helps to stimulate relevant behavior in the female, while her behavior stimulates both sexual behavior and hormonal changes in the male.

Men are quite like rats in regard to the effects of testosterone. The effect of castration of adult men is much more variable than in rats, and reports of the effects do not agree closely (see Shabsigh, 1997). What is clear, however, is that, while many men become effectively asexual within weeks or months, some of these show recovery of interest and function. Others show loss of some function, but retain sexual interest, while still others retain full sexual capacity and interest. Testosterone levels in men also *respond* to sexual stimulation: viewing erotic films increases blood testosterone levels. However, just as in rats, there is generally no relationship between the level of testosterone and the intensity of sexual desire or activity (Sherwin, 1988).

Activating effects in females

The readiness of females of many species to engage in proceptive and receptive behavior depends on cyclic release of gonadal hormones in the estrus cycle. In rats, for example, this cycle is four days, with circulating levels of the estrogen **estradiol** peaking some 40 hours before she becomes receptive. Immediately before the receptive period, progesterone is secreted, coinciding with ovulation. So, sexual behavior in mammals other than primates is virtually restricted to the most fertile phase of the cycle. The consequences of ovariectomy (surgical removal of the ovaries) in female rats and guinea pigs are quick and highly consistent: rapid loss of sexual behavior. The reinstatement of full proceptive and receptive behavior in ovariectomized rats by hormone replacement requires matching the sequence of hormone production. Then, sexual activity rapidly and completely returns (Takahashi, 1990).

Turning to women, the human menstrual cycle involves a cycle of hormonal changes, including the same sequential secretion of estradiol and progesterone as in rats (see *Fig. 1*). Estradiol levels peak around the time of ovulation, and progesterone peaks a few days later. This suggests that women should be sexually most active when most fertile. However, most research has shown that women experience their highest level of sexual desire or activity at around the time of menstruation, usually immediately after, when conception is least likely

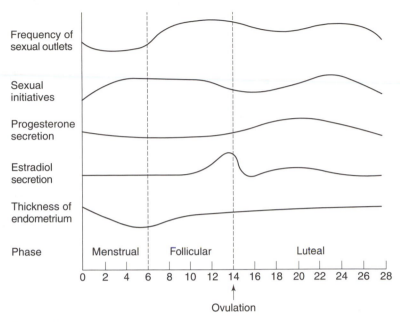

Fig. 1. Changes during the human menstrual cycle.

(see McNeill, 1994). Such a pattern might result from abstinence during menstruation, but the important point is that it suggests that there is no relation between sexual arousability, desire or activity and levels of circulating estrogens or progesterone. However, Baker and Bellis (1995) have shown that women's sexual activity appears to have different immediate aims at different parts of the cycle. Women are more likely to have sex with somebody other than their usual partner during the most fertile part of the cycle, but with their usual partner during the least fertile period.

The effects of ovariectomy and of the loss of female hormones after the menopause also show the lack of close relation between women's sexual behavior and female hormones. Neither of these directly affects sexual interest, nor necessarily behavior. The sexual drive of women is more closely affected by androgens (secreted by the adrenal cortex and the ovaries) than by estrogens. Several lines of evidence show this, including:

- there is a correlation between sexual interest and testosterone in healthy women (Alexander and Sherwin, 1993);
- testosterone injections increase sexual interest in women after ovariectomy and after the menopause, but estradiol injections have no effect (Sherwin and Gelfand, 1987);
- circulating testosterone peaks during the most fertile phase of the menstrual cycle (van Goozen et al., 1995).

Organizing effects of testosterone

The development of male and female sexual behavior depends on the perinatal presence or absence of androgens. To show this, testosterone is injected into a pregnant guinea pig (Phoenix et al., 1959; see *Fig. 2*). Her *female* offspring are tested by removal of the ovaries and subsequent replacement of hormones. These female offspring show a masculine response if injected with androgen;

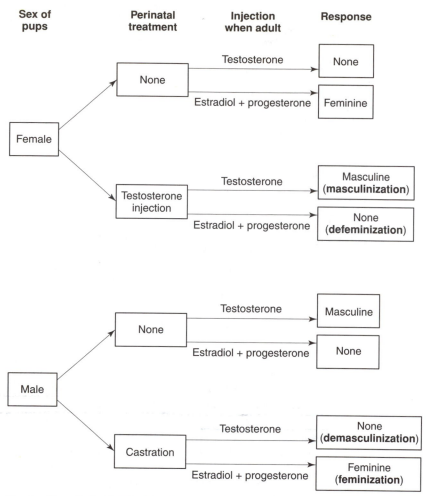

Sex of pups	Perinatal treatment	Injection when adult	Response

Fig. 2.   *The effects of perinatal androgens on later response to hormones. Male sexual behavior develops in either sex under the influence of perinatal androgens.*

that is, they will attempt to mount other females. This is called **masculinization**. Further, they do not show the normal proceptive and receptive responses to estrogen and progesterone injections: **defeminization**. The reverse pattern is shown in male animals. Here, the manipulation is to *prevent* testosterone exposure by castrating the guinea pigs at birth. Later hormone trials show **feminization**, in which the males respond with female patterns of behavior to estrogen and progesterone injections, and **demasculinization**, in which injection of testosterone into the adult males does not produce the usual mounting and copulatory behavior in the presence of a receptive female.

Perinatal testosterone levels thus have sex-specific organizing effects on rodents. Most evidence points to this being an effect on brain structures, and we will look at this in Topic N4. The experiments done on guinea pigs cannot, of course, be carried out on people, but two pathological conditions occur which have similar effects.

Androgen insensitivity syndrome occurs in people who are genetically male, and have testes that produce testosterone and anti-Müllerian hormone.

However, they have no androgen receptors, so the testosterone they produce is ineffective. The AMH inhibits the development of female internal genitalia, and the effective absence of testosterone prevents male genitalia developing. They are born with female external genitalia, and are usually raised as girls. At puberty, the body responds to the small amount of estrogens produced in the adrenal cortex, and what appear to be normal female changes occur (although they do not start menstruating). These people usually think of themselves as women, and show female sexual preferences and behavior. This could be interpreted as feminization and demasculinization of genetic males in the effective absence of perinatal testosterone (Money and Ehrhardt, 1972). But, the female behavior of these people could result from them being treated as female all their lives.

Adrenogenital syndrome (or **congenital adrenal hyperplasia**) occurs both in genetic males and genetic females in whom the adrenal cortex secretes unusually high amounts of androgens, because of genetic errors. In genetic females, the effect varies, depending how much androgen is produced. It results in varying degrees of masculinization of the external genitalia, often giving an indeterminate appearance of gender. Before the cause and treatment of these cases was worked out in around 1950, these genetically female children were raised as either boys or girls, depending on the form of the genitalia. At puberty, there was no problem for those assigned to the female gender, as a normal female puberty ensued. However, for those raised as boys, the female puberty was inconsistent with their assigned gender, and this often caused distress. Genetic females with adrenogenital syndrome are more likely to choose female sexual partners than are the rest of the female population, even those given surgery immediately after birth, and treated for the hormone imbalance (Dittman, Kappes and Kappes, 1992). However, these are unusual people, who are constantly aware of their condition. The outcome might result from altered self-perception, and/or their treatment by other people. Clearly, we cannot conclude that organizational effects are as clear in human as they are in rodents.

N3 NEURAL MECHANISMS IN SEXUAL BEHAVIOR

Key Notes

Male sexual behavior	The medial preoptic area in the anterior hypothalamus is essential for sexual behavior of male laboratory animals, but not for sexual motivation. It receives inputs from neural networks that integrate sensory information, including olfactory and pheromone signals. This circuit is stimulated by hormones.
Female sexual behavior	The anterior part of the ventromedial hypothalamus is important in sexual behavior in female animals. Estrogens cause neurons there to develop progesterone receptors, without which the brain will not respond to progesterone.
Sexual dimorphism in the brain	The sexually dimorphic nucleus in the MPA is much larger in male than female rats and humans. This sexual dimorphism results from an organizing effect of perinatal androgens, and is the locus of action of some activating effects of testosterone. The MPA controls the secretion of hormones that are typical of females and males.
Sexual orientation	There are no differences in hormone levels between gay and straight men. Stress in pregnancy leads to an increased likelihood of homosexuality in offspring in humans and to feminization of behavior in male rodents. Stress increases cortisol production, which reduces testosterone levels. Male homosexuality might result from deandrogenization of the developing brain. Various structural differences that might result from deandrogenization have been reported in the brains of homosexual men, although the significance of the differences is not clear.
Related topic	Hormonal control of sexual behavior (N2)

Male sexual behavior

It has long been known that the **medial preoptic area** (**MPA**) in the anterior hypothalamus is essential for sexual behavior in male laboratory animals. Many lines of research have shown the nature of the control this area exerts over sexual behavior. For example:

- destruction of this area in many species permanently stops sexual activity (Heimer and Larsson, 1966);
- electrical stimulation of this region produces mounting behavior in male rats (Malsbury, 1972);
- activity of neurons in the MPA increases during copulation (Mas, 1995);
- operated male rats show preparatory behaviors in the presence of a receptive female (pursuit and sniffing), but will not mount her;

- operated male monkeys will masturbate, and show interest in a female; however, they will not mount the female.

In summary, the MPA seems to be essential for the programming of male copulatory behavior, but not for sexual motivation. The MPA receives inputs from neural networks that integrate sensory information (including olfactory inputs and pheromone signals from the olfactory bulbs). It also receives inputs from a circuit involving the **medial amygdala** and the **median preoptic nucleus**. This circuit is stimulated by hormones. The MPA is connected to the **lateral tegmental area** in the midbrain, which organizes the motor control of male sexual activity.

The status of **pheromones** in humans is controversial (see Topic I2). Women living together tend to synchronize their menstrual cycles, and some research has suggested that pheromone signals underlie this. Stern and McClintock (1998) reported that sweat from women's armpits placed on the lips of other women could affect the timing of their menstrual cycle. However, this research remains controversial, and some have doubted that menstrual synchrony even occurs (Wilson, 1992). In other species the **vomeronasal organ** in the nose mediates the effects of pheromones. While there is evidence for the existence of a vomeronasal organ in human fetuses, it is not generally considered to be functional in humans (Takami, 2002).

Female sexual behavior

Lesions in the MPA have very little effect on female rodents. However, the anterior part of the **ventromedial hypothalamus** (**VMH**) is important in sexual behavior in female animals. Female rats with VMH lesions do not show receptive behaviors, and may attack a male rat trying to mount them. Electrical stimulation of this area produces lordosis (Pfaff and Sakuma, 1979). The sequentially dependent nature of the female response to hormones results from their actions on cells here. Estrogens cause them to develop progesterone receptors. Without these, the brain will not respond to progesterone, and the female rat will not adopt the receptive position. The progesterone-sensitive neurons in the VMH send fibers to the **periaqueductal gray** area in the brain stem (Hennessey et al., 1990), which is involved in the coordination of species-typical consummatory behaviors.

Sexual dimorphism in the brain

Gorski et al. (1978) and his colleagues demonstrated that the MPA has a nucleus that is three to five times larger in male rats than in female rats. This region became known as the **sexually dimorphic nucleus** (**SDN**) of the preoptic area. Such a difference has also been found in humans. Castration immediately after birth prevents the SDN from retaining its greater size in males, so this sexual dimorphism of the central nervous system is the result of an organizing effect of perinatal androgens (Rhees, Shryne and Gorski, 1990). Castration of adult animals causes some areas within the SDN to shrink, suggesting that this is where some of the activating effects of testosterone take place (Bloch and Gorski, 1988). The MPA controls the cyclical or constant rates of secretion of hormones that are typical of females and males, respectively. A number of other sexually dimorphic areas have been found in the CNS. These include the **third interstitial nucleus of the anterior hypothalamus** (**INAH-3**), which we will look at below, the **bed nucleus of the stria terminalis** (**BNST**; the stria terminalis is a tract that connects the amygdala to the MPA), which is larger in men than in women, and one in the spinal cord that is involved in penile erection.

Sexual
orientation

Exclusive homosexuality is rare in mammals. Its frequency in humans is less than 1% in all societies studied, although the proportion of men showing bisexuality varies from one society to another (Baker and Bellis, 1995).

Some early studies reported lower levels of testosterone in gay men than in straight. However, later research has generally failed to confirm this, and the earlier results might have reflected greater stress in the gay subjects (Meyer-Bahlburg, 1984). One effect of stress is to increase **cortisol** secretion (see Topic O4), which in turn suppresses testosterone. Furthermore, as we have seen, androgen levels are important in sexual desire and behavior in men *and* in women, and there is no reason to suppose they should be lower in gay men.

Even if activating effects of hormones do not differ between straight and gay people, organizing effects might do. Evidence from the genetic androgen deficiencies mentioned earlier does not clearly indicate a biological rather than a socio–cultural explanation for sexuality. In the 1980s Dorner reported that in humans stress in pregnancy leads to an increased likelihood of homosexuality in offspring (e.g. Dorner et al., 1983). It was already known that in pregnant rats stress increases cortisol production, and leads to feminization of behavior of male offspring. This suggests that male homosexuality might result from **de-androgenization** of the developing brain. Are there differences in the brains of heterosexual and homosexual people?

Swaab and Hofman (1990) found that the **suprachiasmatic nucleus** of homosexual men is about twice the size it is in heterosexual men. Swaab et al. (1995) demonstrated that preventing perinatal testosterone acting on the brains of developing rats produces an imbalance of neuron types, and causes the rats to be bi-sexual in choice of sexual partner. LeVay (1991) found that the INAH-3, which we saw earlier is on average twice as big in men as in women, is the same size in gay men as in women. However, there is an enormous amount of overlap in the two male distributions. Furthermore, the difference might be a result of early experiences, or be a *result* of the sexual preference. The sexual dimorphism of the INAH-3 is dependent on perinatal testosterone levels and so, although its role in sexual behavior is not known, it could be the location of a deandrogenization basis for homosexuality. The result appears still not to be replicated.

Homosexual men are also reported to have a larger **anterior commissure** (a fiber bundle connecting the two cerebral cortices) than straight men, and about the same size as that of women (Allen and Gorski, 1992). The sex difference in this structure is usually thought to relate to cognitive differences rather than differences in sexual behavior. This is true also of differences in EEG lateralization that have been observed between gay and straight men. Again, task-related EEG patterns in homosexual men are closer to those of women.

A recent review of research on brain structures and sexual orientation (Mbugua, 2003) has concluded that the research has so many conceptual and methodological flaws that we cannot be sure that there are structural differences in the brains of gay and straight men.

N4 PARENTAL BEHAVIOUR

Key Notes

Hormonal control of parental behavior	All mammalian infants depend on parents, primarily the mother. In laboratory rats, males and unmated females will not normally care for pups. If left in contact with pups for several days unmated females will start to care for them. This sensitization and the avoidance depend on odor. Progesterone, prolactin and the sequence of high progesterone followed by high estrogen sensitize aspects of maternal behavior. A nursing rodent shows maternal aggression towards males or females who come close. For two days after birth aggression is absent. The female is likely to mate at this time. The return of the aggressive behavior depends on the infants suckling from the mother.
Neural mechanisms in parental behavior	The medial preoptic area controls maternal behavior through neurons in the brain stem. The MPA receives inputs from the medial amygdala. The medial amygdala receives information from both olfactory systems. Estrogen has its facilitatory effect on maternal behavior through estrogen receptors in the MPA, which have to be primed by prior exposure to progesterone.
Related topics	Reproduction and sexual differentiation (N1) Aggression (O4)

Hormonal control of parental behavior

The different mating patterns that we saw in Topic N1 lead to different patterns of parental behavior. All mammalian infants are dependent on one or both parents (or extended family group or tribe) for protection. They are also all dependent on their mothers for sustenance. Even when the father participates in the care of the infant, the dependency is primarily, or at least initially, on the mother. The mother's care not only involves providing nutrition, but also uses species-typical behaviors that assist, for example, maintenance of the infant's body temperature, and stimulation of the development of important functions such as standing and locomotion, defecation and passing of urine.

In laboratory rats, males and unmated females will not normally care for pups. Indeed, unmated females avoid young pups. However, if left in contact with pups for several days they will start to care for them. This is called **sensitization**. Once a rodent is sensitized in this way, she will in future always try to care for pups she encounters. Disrupting either the olfactory system or the accessory olfactory system (responsible for pheromone detection) prevents the avoidance by unmated females, and leads them to care for pups (Fleming and Rosenblatt, 1974). This shows that the avoidance is based on odor, including pheromones. Mothers will also respond to specific calls emitted by pups which indicate that the pups are, for example, cold or hungry. The fact that these maternal responses follow pregnancy suggests that hormonal changes sensitize maternal behavior.

Levels of circulating hormones vary dramatically during pregnancy (see *Fig. 1*). Experimental studies in rats have shown that these changes are related

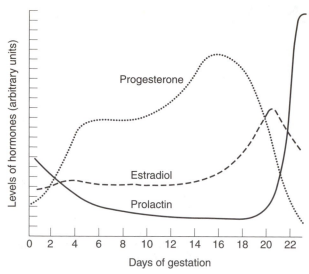

Fig. 1.   Levels of three hormones during pregnancy in the rat.

to aspects of maternal behavior. For example, administering **progesterone** to unmated rats stimulates nest building (Lisk, Pretlow and Friedman, 1969). This mimics the high level of progesterone through most of the second half of pregnancy, which is when nest building normally occurs. But progesterone cannot be the only factor, since nest building continues after the birth of the pups (**parturition**), even though progesterone levels are by then very low. At this stage, **prolactin** (an anterior pituitary hormone which is very high just before parturition, and which stimulates lactation) is important. Administration of prolactin also stimulates nest building (Bridges et al., 1990). The posterior pituitary hormone **oxytocin**, which stimulates milk release during suckling, also promotes parenting behavior in various species (e.g. Gonzalez-Mariscal, 2001).

Other aspects of maternal behavior are stimulated by the sequence of high progesterone followed by high **estrogen**. Administering progesterone followed by **estrogen** facilitates maternal behavior in unmated female rats (compare the effects of these hormones on female sexual behavior, Topic N2). The result is like sensitization, but occurs more quickly (e.g. Fleming et al., 1989).

In primates, including humans, evidence for an endocrine influence on maternal behavior has only recently been produced (e.g. Pryce, 1993). Levels of **cortisol** are higher in women who are more attracted to their babies' scents, and they show greater bonding with the infants (Fleming, Steiner and Corter, 1997). However, the hormone effects are small compared with parental learning and environment. Another aspect of maternal behavior is **maternal aggression** (see Topic O4). A nursing mother rat or mouse will instantly attack males or females who come close to their pups. Like other aspects of maternal behavior, maternal aggression starts before she gives birth, as progesterone levels rise. Immediately after birth however, and for about two days, aggression is almost completely absent. The female is receptive at this time and is likely to mate if a male approaches. This phenomenon is probably caused by estradiol, since removal of the ovaries maintains aggression after parturition, and injection of estradiol stops it. The return of the aggressive behavior depends not only on the presence of infants, but on the infants suckling from the mother. Removal of the nipples prevents the return of aggression, and this does not depend on hormones from the ovaries (Svare et al., 1982).

Neural mechanisms in parental behavior

Lesions of the forebrain **medial preoptic area** (**MPA**) completely abolish all aspects of maternal behavior (Numan, 1974). (The MPA is crucial for male sexual behavior, but plays no part in female sexual behavior; see Topic N3.) Maternal behavior is organized by neurons in the brain stem, which are connected to the MPA by pathways through the **ventral tegmental** area. The MPA receives inputs from the **medial amygdala** by way of the **stria terminalis**. Destruction of either the stria terminalis or the medial amygdala stops unmated rats having an aversion for pups (Fleming, Vaccarino and Luebke, 1980). The medial amygdala, in turn, receives information from both olfactory systems and from other senses. Thus, the amygdala has an inhibitory effect on the MPA, and this is triggered by stimuli like the odor of pups. Estrogen has its facilitatory effect on maternal behavior through estrogen receptors in the MPA, which have to be primed by prior exposure to progesterone. While the neural basis of sensitization is not clear, estrogens seem to facilitate maternal behavior by reducing the inhibitory effect of the amygdala.

Where maternal aggression is concerned, sensory and hormonal stimuli discussed above combine with stimuli from an intruder to stimulate probably the same neural circuits as those of affective aggression in general (Lonstein and Gammie, 2002; see Topic O4).

01 PERIPHERAL FACTORS IN EMOTION

Key Notes

The James–Lange theory of emotion	The James–Lange theory of emotion holds that emotions result from our perception of physiological changes resulting from emotional stimuli. Cannon attacked the James–Lange theory on five points. Separation of the viscera from the central nervous system does not alter emotional behavior. The same visceral changes occur in many states. Induction of the bodily changes of emotions does not produce them. Viscera are relatively insensitive. Visceral changes are too slow. Cannon's critique is wrong in many respects. The viscera are not insensitive; they are richly endowed with afferents.
Spinal cord lesions	Early studies of patients with spinal cord lesions provided evidence that the lesions result in decreased intensity of emotional feeling. Some suggested that the decrement is greater with lesions higher up the spinal cord. Later studies have failed to find any such effects of injury.
Physiological differentiation of emotions	Studies of physiological changes in different emotions have led to the conclusion that anger is associated with the excretion of both epinephrine and norepinephrine, and fear with the increased production of epinephrine. There is very little evidence for differentiation of other emotions. There is little good evidence that feedback from facial muscles determines emotional feelings.
Effects of catecholamine infusion	Injections of catecholamines generally produce anxiety or 'as-if' emotions: physiological sensations without emotional significance. When real emotions are reported these might be attributed to physiological activation of pre-existing moods. Schachter's two-factor theory of emotion claimed that emotion results from physiological arousal labeled by cognitive appraisal. The data do not support the theory.
Related topic	Central mechanisms in emotion (O2)

The James–Lange theory of emotion It is widely agreed that emotions have subjective, behavioral, cognitive and peripheral physiological components. From a common sense viewpoint, the physiological and expressive changes seem to be the *results of* emotional arousal. However, William James (1884) proposed that emotions *arise from* the physiological (and behavioral) changes that result from the perception of an emotional stimulus. For instance, 'we feel sorry *because* we cry; angry *because* we strike; afraid *because* we tremble' (see *Fig. 1*). James argued that trying to imagine a strong emotion without its characteristic bodily symptoms leaves a cold, intellectual perception. Conversely, imagining the bodily changes of an emotion can lead to the emotion itself, and adopting opposing bodily postures and

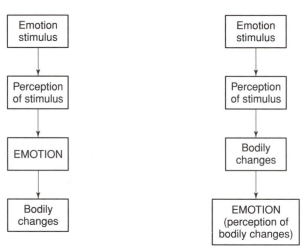

COMMON SENSE VIEW

William James (1884):

"We lose our fortune, are sorry and weep; we meet a bear, are frightened and run; we are insulted by a rival, are angry and strike."

JAMES–LANGE THEORY

*"We feel sorry **because** we cry; angry **because** we strike; afraid **because** we tremble."*

Fig. 1.   The common sense and James–Lange theories of emotion.

actions can dispel the emotion. The following year Carl Lange proposed a theory that attributed emotions to the vasomotor system (which controls the circulation). The result was the dominant theory of emotion for the next 30 years or so: **the James–Lange theory**. The essentials of the theory are that different emotions have different patterns of physiological response, and that reducing the afferent feedback of these responses should reduce the intensity of the emotion.

Walter Cannon (1927) criticized the James–Lange theory with a 5-point attack:

- *total separation of the viscera from the central nervous system does not alter emotional behavior*. Experiments on dogs and cats by Sherrington (1900) and Cannon et al. (1929) showed that the animals' emotional responses were apparently unaffected by neural lesions that almost completely removed feedback from the viscera;
- *the same visceral changes occur in different emotional states and in non-emotional states*. Cannon's (1929) own work on physiological changes in hunger, pain, fear and anger demonstrated what seemed to be general sympathetic arousal in all of these states (the fight-or-flight response);
- *artificial induction of the visceral changes typical of strong emotions does not produce them*. Studies of the effect of injection of adrenal extracts into humans apparently do not produce reports of emotions;
- *the viscera are relatively insensitive structures*. That is, we have only limited awareness of changes in the viscera;
- *visceral changes are too slow to be a source of emotional feeling*. Cannon claimed that affective reactions end in under a second, while many vascular and glandular responses take several seconds.

It gradually became clear that Cannon's critique was wrong in many respects. The viscera are not insensitive; they are richly endowed with afferents, without which homeostasis could not occur (see Topic L1), although we are not necessarily *aware* of this afferent input. Voluntary muscle responses are fast enough to generate emotions. In the following sections we look at evidence concerning the role of peripheral physiological changes in emotion.

Spinal cord lesions

Sherrington's (1900) and Cannon et al.'s (1929) studies of reduced sensory feedback are not convincing for two reasons:

- the animals could have responded through conditioned responses, rather than through emotional *experience*;
- the procedures left some afferent pathways intact.

A number of studies have looked at the effects of reduced peripheral feedback in humans. If emotional feeling results from feedback from peripheral physiological activity, a complete lesion of the spinal cord would lead to decreased intensity of emotional feeling. These decrements would be greater for patients with lesions higher up the spinal cord, since they have greater reduction in afferent feedback. Dana (1921) examined a woman with a high spinal fracture at the level of the 3rd and 4th cervical vertebrae (in the neck). She showed a range of emotions. Hohmann (1966) interviewed paraplegic and quadriplegic patients with lesions at five levels: cervical, upper thoracic, lower thoracic, lumbar and sacral. Overall, they reported a decrease in intensity of several emotions, with a tendency for the decrease to be greater the higher the lesion.

In contrast, a number of studies conducted in the past 20 years have shown that the majority of spinal cord patients report either no change or *increases* in emotional feelings after injury (e.g. Lowe and Carroll, 1985; Bermond et al., 1991). Further, they generally found no relationships with level of lesion. These results are inconsistent with James's theory. Hohmann's findings might have been due to the more pessimistic approach to spinal cord injury in the mid 20th century.

Physiological differentiation of emotions

The endocrine and autonomic nervous systems are capable of much more differentiated activity than Cannon was aware of (see Jänig, 2003). If emotions prepare us for particular types of action we should expect them to have different physiological patterns. Some physiological differentiation is obvious. For example, we feel nausea in disgust and fear, 'butterflies' when nervous, the blood rushing in our ears when angry, and a 'lump in the throat' when sad; our faces may burn when we are embarrassed, our hearts pound when afraid, our palms perspire when afraid, and we blush with embarrassment and blanch with fear.

Formal studies to reveal physiological differentiation are fraught with difficulties (see Wagner, 1989). Not only is it necessary to produce *pure* emotions, but also to measure appropriate physiological parameters, at the appropriate time. The first may be near impossible, and studies have generally just measured the parameters they happen to have the equipment for. Ax (1953), for example, induced fear by simulating a short circuit in the electrophysiological equipment to which the subject was attached. Anger was induced by an assistant loudly blaming the subject for having produced the short circuit. In general, *anger* is associated with the excretion of both epinephrine and norepinephrine, and *fear* with the increased production of epinephrine. There is very little evidence for differentiation of other emotions, although this might be due to the difficulty of such research.

Another way emotions might be differentiated is by way of the muscles of facial expression (see Topic O3). **Facial feedback theories** of emotion argue that our emotional experience is determined by the expression we have on our faces (see Laird, 1974). Many studies have been conducted to examine this view, but most of these suffer from methodological inadequacies. At most, it has been shown that facial expressions might *accentuate* an emotional response (if appropriate to the emotion), or diminish it (if inappropriate). However, there is no incontrovertible evidence that they can themselves produce emotions. What effects have been demonstrated are small in comparison to the effects of external stimuli. Keillor et al. (2002) reported a case of a woman with complete paralysis of the facial muscles who nevertheless responded normally to emotional stimuli.

Effects of catecholamine infusion

Cannon supported his claim that artificial induction of the visceral changes typical of strong emotions does not produce them, by reference to the work of Marañon (1924). Marañon injected people with adrenaline (adrenal medullary extracts; a mixture of epinephrine and norepinephrine) and recorded their responses. Most reported 'cold, as-if' emotions, mostly of fear, with physical symptoms but no mental component. He concluded that the general sympathetic activation induced by adrenaline produces physical sensations but not true emotion. Later studies infusing adrenaline, or epinephrine and norepinephrine separately, have similarly reported preponderance of somatic changes. When what appear to be 'real' emotions are reported these might be attributed to physiological activation 'energizing' the anxiety experienced in the laboratory or pre-existing moods (Breggin, 1964).

Schachter (1964) elaborated this view into a physiological and cognitive **two-factor theory of emotion**. He claimed that emotion only results when otherwise-unexplained *physiological arousal* is labeled by *cognitive appraisal*. Schachter and Singer (1962) tried to manipulate these two factors separately by generating unexplained arousal with epinephrine injections, and cognitive appraisal by pairing each participant with a confederate who behaved angrily or happily. They claimed that the results showed that aroused participants could be 'readily manipulated' into feeling anger or euphoria. In fact, the results of the study did not demonstrate that at all. Both groups of participants reported negative states, even though one group was supposed to be experiencing a euphoric state (Manstead and Wagner, 1981; see *Fig. 2*). Despite the enormous influence the theory has had, attempts by others to support it have also been unsuccessful.

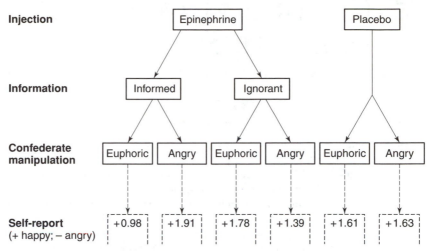

Fig. 2.   The main results of the Schachter–Singer study. They predicted that participants ignorant of the arousing effects of the epinephrine injection would adopt the emotional state of the confederate. The results show that all groups reported themselves to be on the euphoric end of the mood measure.

02 CENTRAL MECHANISMS IN EMOTION

Key Notes

Cannon–Bard theory	The Cannon–Bard theory proposes that thalamic and hypothalamic centers organize emotional responses to stimuli. These centers are controlled by inhibition from the cerebral cortex. Removal of parts of the temporal lobes and the limbic system produces the Klüver–Bucy syndrome: docility, hypersexuality and indiscriminate eating. The Papez circuit is a 'neural circuit of emotions' in the limbic system. Lesions in any part of this circuit are associated with emotional abnormalities.
Emotion and the amygdala	The amygdala is in the anterior part of the temporal lobe. Its nuclei mediate the effects of sensory stimuli on emotional behavior and through interconnections with other cortical and subcortical structures it controls emotional expression and behavior. Much of the Klüver–Bucy syndrome results from removal of the amygdala. The amygdala is especially important in the processing and learning of fear-related stimuli.
The orbitofrontal cortex	The frontal lobes, especially the orbitofrontal cortex, receive inputs from various brain locations involved in emotion. It sends fibers to the hippocampus, the amygdala, and the lateral hypothalamus, amongst other structures. Destruction of the frontal lobes has a calming effect. This led to the development of lobotomy, cutting the fiber tracts connecting the frontal lobes with other parts of the brain, as a treatment for anxiety, depression, and obsessive-compulsive behavior. The orbitofrontal cortex plays a role in reward mechanisms, and seems to mediate the of use emotion to direct judgments.
The cingulate cortex	The cingulate cortex is activated by pain, with different parts activated by different aspects of pain. It is also activated in depression.
Hemispheric lateralization and emotion	Data currently support two different views of the emotional roles of the two cerebral hemispheres. One is that the right hemisphere is specialized for, amongst other things, emotional functions and the left hemisphere for verbal activities. The other is that the left hemisphere is specialized for positive, and the right hemisphere for negative emotion.
Related topics	Peripheral mechanisms in emotion (O1) Aggression (O3)

Cannon–Bard theory

Following his attack on the James–Lange peripheral theory of emotion (see Topic O1), Cannon proposed a central theory; later elaborated by Bard into the **Cannon–Bard theory**. Bard (1928) showed that removal of the cerebral cortex of cats may result in extreme, undirected aggressive behavior: **sham rage**.

Subsequent removal of the **hypothalamus** or the lower posterior portion of the **thalamus** stops the sham rage. These studies are consistent with observations of humans. The lighter stages of general anesthesia suppress cortical processes, but not subcortical structures. With slow acting anesthetics like nitrous oxide, a patient remains in this state long enough for it to produce emotional responses (hence the name 'laughing gas'). Cannon proposed that thalamic and hypothalamic centers organize sympathetic and expressive responses to stimuli, and give rise to emotional experience. These centers are controlled by the cerebral cortex through variations in inhibition following the perception of emotional objects (see *Fig. 1*).

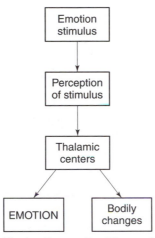

Fig. 1.   Cannon–Bard theory of emotion. (Compare Fig. 1, Topic O1.)

Klüver and Bucy (1938) found that removal of parts of the temporal lobes of monkeys caused them to lose their fear and aggression, and become docile. They also showed hypersexuality, and ate indiscriminately. This became known as the **Klüver–Bucy syndrome**. It was later discovered that this syndrome was only observed if the lesions included subcortical structures in the **limbic system**; a group of interconnected centers in the cerebral hemispheres and diencephalon, surrounding the thalamus (see *Fig. 2*; but note that there is still uncertainty about what structures are part of the limbic system). Papez (1937) after examining pathological brains from humans and animals, described a 'neural circuit of emotions' (the **Papez** circuit) in the limbic system. Lesions in any part of this circuit were associated with emotional abnormalities. On the basis of this Papez concluded that the physiological changes in emotion arise from hypothalamic centers controlling visceral organs, while feelings arise (in some way not clearly specified) in the circuit as a whole, including the **mammillary bodies**, **anterior thalamus** and **cingulate cortex**. Later work, using stimulation and lesion methods, has added the **amygdala** and the **septum**.

Emotion and the amygdala

The **amygdala** is located in the anterior part of the temporal lobe. It consists of a number of groups of nuclei. These include:

- the **medial nucleus**, which receives sensory inputs, and mediates the effects of, for example, odors on sexual behavior (see Topic N2) and aggression (see Topic O4);

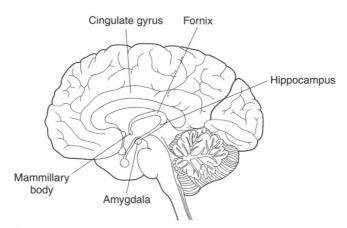

Fig. 2.   The main components of the limbic system.

- the **lateral** and **basolateral nuclei**, which receive information from areas of the cerebral cortex, and from other parts of the limbic system. They connect with other centers, including:
- the **central nucleus**, which in turn connects to the hypothalamus and other subcortical structures that control components of emotional expression and behavior.

Many of the effects of destruction of the temporal lobe (the Klüver–Bucy syndrome) result from removal of the amygdala. Through its various connections, and because it is activated by hormones from both the adrenal medulla and cortex (see Topics E3 and O4), the amygdala plays a central role in the organization of emotional experience and behavior, and the arousal of emotion by sensory stimuli.

The best understood emotional role of the amygdala is in fear. Destruction of the amygdala in many species produces insensitivity to fear-related stimuli (see LeDoux, 1995). The central nucleus is particularly important in the mediation of fear responses. Stimuli that are conditioned to produce fear responses (for example, by pairing them with electric shocks) do not evoke a response in neurons in the central nucleus before conditioning, but do so afterwards. Lesions in the central nucleus prevent such conditioned fear responses. Stimulation of the central nucleus produces autonomic signs of arousal, as well as behavior associated with vigilance.

In humans, too, lesions of the amygdala impair the ability to learn emotional responses. Furthermore, electrical stimulation of the amygdala produces reports of fear or anxiety, while stimulation of other limbic system structures produces, at most, autonomic components of an emotional response. Studies using PET scans (see Topic A2) have shown that activity in the amygdala increases during recall of, or exposure to, various types of emotional stimuli, including threatening words (e.g. Isenberg et al., 1999). The amygdala is not only involved in conditioned fear, it is also involved in the effects of emotion on learning. People more readily remember events associated with emotional reactions. However, a patient with bilateral degeneration of the amygdalas showed no better recall of emotional than unemotional material (Cahill et al., 1995).

The orbitofrontal cortex

We have seen that in the 1930s the cerebral cortex was shown to have a generally inhibitory influence on emotional behavior. The frontal lobes of the cerebral cortex, in particular the **orbitofrontal cortex** at the base of the frontal lobes, receive inputs from various brain locations involved in emotion. These include the amygdala, the thalamus, the ventral tegmental area and the olfactory system. In turn it sends fibers to the hippocampus, the amygdala and the lateral hypothalamus, amongst other structures. So it is well placed to play a central role in emotion. Its role was brought to light in the mid-19th century, following the famous case of Phineas Gage. In an accident with dynamite, a steel rod was blown through his cheek and the frontal regions of his brain, and out through the top of his skull. Before the accident, he had been a serious, efficient and energetic worker. Afterwards, he gradually became childish, apathetic, rude, irresponsible and thoughtless. He was also unable to make or follow through plans. Examination of his skull suggests that there was considerable damage to his frontal lobes, particularly to the left orbitofrontal cortex. Later observations of patients with injuries to the orbitofrontal cortex confirmed the general picture of reduction in inhibition and lack of concern for the effects of their behavior.

Jacobsen et al. (1935) showed that destruction of the frontal lobes in a chimpanzee that showed violent reactions calmed the animal. At the same time, it was reported that the removal of the frontal lobes to treat a tumor in a human patient did not affect his intelligence. This led to the development of **lobotomy**, the treatment of psychiatric disorders by cutting the fiber tracts connecting the frontal lobes with other parts of the brain. The emotional consequences of this surgery were to reduce anxiety, depression and obsessive-compulsive behavior. However, lobotomy does have a general emotional flattening effect, and causes difficulty with social interaction, as well as with planning, so that most patients are unemployable.

The precise role played by the orbitofrontal cortex in emotion is still unclear. Brain imaging studies have shown that it is involved in response to pleasant and unpleasant situations, as varied as touch and odors (Francis et al., 1999), food (see Topic M3) and monetary reward (Thut et al., 1997). It seems, therefore, to be part of the brain's reward system (Topic P4). However, it appears to play a role also in the use of sensory and physiological 'markers' of emotional activation to guide behavior. Although people with destruction of this area can deal with hypothetical situations involving moral or ethical dilemmas, they appear to be unable to apply the same principles in their own interactions. It seems that in real life such judgments are made with the mediation of emotional processing, and that this is absent in such patients. The role of the orbitofrontal cortex seems to be to use emotion to direct judgments (Damasio, 1996).

The cingulate cortex

The **cingulate cortex** is a region of the cerebral cortex surrounding the thalamus, and has been described as part of the Papez circuit (see above). Imaging studies in humans show that this region is activated by pain or the anticipation of pain. Tolle et al. (1999) have shown that different aspects of pain (sensory, affective and gating; see Topic I1) activate different parts of the cingulate cortex, as well as other brain regions. Surgical destruction of the cingulate cortex has been used to treat intractable pain in humans (Kondziolka, 1999). Activation of the cingulate cortex has also been shown in clinical depression, and even in sad moods (Mayberg et al., 1999).

Hemispheric lateralization and emotion

It has been argued that the two cerebral hemispheres have different emotional roles (see Tucker and Frederick, 1989). Most research, including observations of 'split-brain' patients (see Topic C3) and patients with unilateral brain lesions, as well as experiments presenting stimuli only to one hemisphere, suggests that the right hemisphere is specialized for emotional functions (together with, e.g. spatial activities), and the left hemisphere for verbal activities (see also Topic Q2). For example, Etcoff (1984) showed that patients with right hemisphere lesions have impaired ability to recognize facial expressions of emotion (see Topic O3).

A different view of hemispheric lateralization has been proposed by Davidson and his colleagues (e.g. Davidson, 1992). They argue that the left hemisphere is specialized for positive affect, and the right hemisphere for negative emotion. This view has its origins in the observation that brain damage to the left hemisphere is more likely to cause deep depression than is damage to the right hemisphere (e.g. Goldstein, 1952). Support comes from numerous experimental sources. For example, intact humans are better able to recognize positive expressions shown to the left than to the right hemisphere (Davidson et al., 1987), and negative stimuli produce greater activation of the right hemisphere (Zald et al., 1998).

O3 EMOTIONAL EXPRESSION

Key Notes

Basic emotions	Many believe that certain emotions are basic, most often because a few emotions seem to have facial expressions that are universally recognizable. Evidence for universality of expressions came from research showing a high level of agreement between different cultures on what emotion is expressed in certain faces. However, the methodology of this work has been criticized. An alternative view, behavioral ecology, is that facial behaviors are displays that have evolved in relation to specific situational pressures; their meaning can only be specified in relation to the context.
Neural mechanisms of emotion expression	Voluntary facial paresis results from lesions in the area of the primary motor cortex or its efferents. Following damage to areas involved in emotion control an individual may be able to produce voluntary expressions, but will not show facial expressions when emotionally aroused. Facial expressions, especially voluntary ones, tend to be stronger in the left side of the face, suggesting right hemisphere specialization for emotional expression.
Neural mechanisms in emotion recognition	The right cerebral hemisphere is especially involved in emotion recognition, including stimuli in different modalities. The amygdala is also involved in emotion recognition.
Related topics	Peripheral factors in emotion (O1) Central mechanisms in emotion (O2)

Basic emotions

A pervasive view in the theory of emotion is that certain emotions are basic (or fundamental, or primary; see Ekman, 1992). There are many reasons for such a proposal:

- since the English language has 200–300 words for states that might be considered emotions, it simplifies the area of study to consider some of them basic;
- some emotions seem to be psychologically irreducible, while others may be analyzed introspectively into simpler emotions. The irreducible emotions may be considered basic;
- some emotions seem to have evolved to perform important tasks related to individual or species survival, so must be innate and hard-wired, and in that sense basic;
- the most frequently cited reason for proposing basic emotions is the apparent demonstration that a few emotions have facial expressions that are universally recognizable. This being so, those emotions must have an evolved, biological basis.

Darwin (1872/1998) argued that human facial expressions are part of innate patterns of response evolved from behaviors that we can still see in other species. These behaviors are habitual responses or 'direct actions of the nervous system' that result from particular eliciting conditions. Over time, these became separated from the original cause and came to express emotional states. Thus, for example, the baring of the teeth we sometimes see in rage is said to be related to the snarl of a wolf, which precedes (or shows the intention to) attack. Darwin supported the supposed innateness and phylogenetic continuity of these behaviors largely with anecdotal evidence about the expression of emotion in animals, children and different racial groups. The dominant view until the second half of the 20th century, however, was that expression was culture-specific and learned (see Birdwhistell, 1970).

More formal evidence for universality of expressions came from research on the recognizability of facial expressions (see Ekman, 1992). In most studies, photographs of white Americans posing the expressions of proposed basic emotions were shown to people in various other cultures, including some considered not to have been exposed to western culture. Usually, a high level of agreement on what emotion is expressed in each face was found. Most researchers consider these results to demonstrate the universality, and hence basic nature, of these emotions. The most frequently accepted list of basic emotions demonstrated in this way is that of Ekman: *anger, contempt, disgust, fear, happiness, sadness* and *surprise* (see *Fig. 1*). However, the methodology of this work has been criticized on various grounds. For example:

• the studies usually force people to identify facial expressions as emotions. However, given a completely free choice of vocabulary to describe expressions, emotion words are rarely used. Most often, people describe situations;
• all of the studies have used *posed* facial expressions, or facial expressions constructed from facial movements, which may or may not be the expressions that occur spontaneously;
• most studies have used *forced-choice* methods, in which participants choose one of a fixed list of emotion labels, without the option of responding 'none of these'. This means that some expressions might be 'correctly' identified by default, because none of the other labels fit.

More fundamentally, the idea that facial (and other) behaviors *express* emotion in the sense of being a 'read-out' of emotional state, or an essential

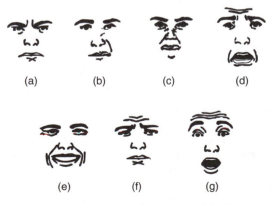

(a)   (b)   (c)   (d)

(e)   (f)   (g)

Fig. 1.   Typical expressions of seven supposed 'basic' emotions: (a) anger; (b) contempt; (c) disgust; (d) fear; (e) happiness; (f) sadness; (g) surprise.

component of it, has been attacked (Fridlund, 1994). There is surprisingly little direct evidence that emotional experience and facial expression coincide. All agree that people can manage their facial behavior. This can be, for example, in response to social and cultural demands (including what are called **display rules**; internalized, culture-specific descriptions of *who* may show *what* emotion to *whom* and *when*; Ekman and Friesen, 1969) or to falsify impressions. An alternative view, **behavioral ecology**, championed by Fridlund (1994), is that facial behaviors are displays that have evolved in relation to specific situational pressures, and their meaning can only be specified in relation to the context.

Neural mechanisms of emotion expression

Even if facial behavior does not express emotions, it is nevertheless true that consistent patterns of facial behavior are produced, and that these are consistently interpreted by others. Some of the neural mechanisms underlying the production and recognition of emotional expressions have been described (Rinn, 1984).

Automatic emotional expressions are often different from those we produce to pretend an emotion. They are more likely to be symmetrical, and they might also involve different muscles. A clear example of this is smiling. The 'Duchenne smile' (first described by Duchenne in 1862) involves contraction of the *zygomaticus* muscles, pulling the corners of the mouth upwards and outwards, and in addition contraction of a muscle surrounding the eyes, the *orbicularis oculi*, which has the effect of lifting the lower eyelid. Most people are unable to contract the *orbicularis oculi* muscles voluntarily, so fake smiles usually only involve the *zygomaticus*. (This does not mean that the Duchenne smile necessarily signifies happiness; simply that it is not produced voluntarily.)

The distinct neural pathways underlying voluntary and spontaneous expressions are revealed by two neurological disorders (Hopf et al., 1992). Lesions in the area of the **primary motor cortex** controlling the face, or in the pathways that connect this to the facial nerve nucleus (the **corticospinal pathway**), cause the sufferer to be unable to voluntarily control the opposite side of the face. When instructed to smile, for example, the person will produce a unilateral expression on the same side of the face as the lesion. However, when happy or amused the person will show a normal, bilateral smile. This is sometimes called **voluntary facial paresis**. Conversely, following damage to part of the **prefrontal cortex**, parts of the **thalamus**, or the **basal ganglia**, an individual may be able to produce voluntary expressions, but will not show facial expressions when emotionally aroused. This sometimes happens in advanced **Parkinson's disease**, when subcortical centers are damaged. These findings show that voluntary expressions are produced in exactly the same way as any other voluntary movement. Spontaneous expressions, though, are produced by the cortical and subcortical circuits that produce emotional behavior and responses (see Topic O2).

Neural mechanisms in emotion recognition

Neural processes in the recognition of emotional expressions are less well understood than those of their production. Experimental data and studies of brain-damaged patients suggest the right cerebral hemisphere is especially involved in the recognition of emotion in facial and vocal emotion (**prosody**; but see Topic O2). For example, Adolphs et al. (2000) showed that judgment of emotions in facial expressions was most affected by lesions in the right somatosensory cortexes (see Topic I1), with different emotions affected by different, but overlapping, regions. Studies with PET scans (see Topic A2) have shown that, while recognition of emotion from the *meaning* of a word produces greater activity in the left, language, hemisphere (see Topic Q2), emotion recognition from *tone of voice* produces greater activity in the right prefrontal cortex (George et al., 1996).

Lesion studies have shown that the amygdala is involved in recognition of emotion, particularly fear, from visual stimuli (e.g. Sato et al., 2002) but perhaps not from prosody (Adolphs and Tranel, 1999). Adolphs, Baron-Cohen and Tranel (2002) have recently suggested that damage to the amygdala produces deficits specifically in relation to social emotions (e.g. guilt) rather than basic emotions (e.g. happiness and anger).

04 AGGRESSION

Key Notes

The nature of aggression

Affective aggression (defensive rage) is aggressive behavior when attacked or threatened, or as a display. It is accompanied by a high level of sympathetic activation. Predation is aggressive behavior directed at a member of another species, usually for food. Competition for territory, mating rights, food etc. produces inter-male aggression. Female aggression is usually deployed in the defense of the young.

Hormones and aggression

Androgens have similar activating effects on offensive aggression as on sexual behavior. They also have organizing effects. Aggression in male rats is affected by social and situational factors, interacting with sex hormones. Although offensive aggression is rare in female rodents, it is influenced by the activating and organizing effects of androgens. Individual differences result from differing exposure to androgens as fetuses. Most female offensive aggression is maternal aggression.

Neural mechanisms of aggression

The motor pattern of affective aggression is programmed by neurons in the periaqueductal gray area. These neurons receive inputs from the medial hypothalamus. The medial hypothalamus normally plays a role in determining the occurrence of aggressive behavior. The amygdala modulates aggressive behavior through inhibitory and excitatory pathways to the PAG. The cerebral cortex exerts an inhibitory control over aggression.

Human aggression

In primates, the early social environment is the main influence on aggressive behavior. Aggression is related to activating effects of testosterone. Testosterone might be related to dominance rather than aggression. The inhibitory effect of the cerebral cortex on aggression in humans is shown by the fact that a number of disorders of the cortex are associated with increased aggressiveness. Alcohol increases aggressiveness by interfering with the executive functions of the prefrontal cortex, which normally control it.

Related topic

Hormonal control of sexual
    behavior (N2)

The nature of aggression

Different types of aggression have been described, but these are not all clearly distinct in different species. For our purposes we will distinguish two:

- **affective aggression** (or **defensive rage**) is aggressive behavior when attacked or threatened, or as a display. It is accompanied by a high level of sympathetic activation, and the animal usually adopts species-typical behavior and postures that make it look more dangerous. Each usually involves multiple or prolonged threat displays, and more than one attack. The end result is usually the withdrawal of one or other of the combatants, often before the confrontation has escalated from threat to actual attack.

- **predation**: aggressive behavior directed at a member of another species, usually for food. The animal makes itself as unobtrusive as possible, and the attack is directed at killing the victim, and usually consists of a single attack, without threat displays. It is not accompanied by high sympathetic arousal.

This distinction is clear in cats (Siegel et al., 1999), and is paralleled in humans (Vitiello et al., 1990). We will concentrate on affective aggression. While aggressive behavior seems to be the same in male and female animals, the nature of aggression does differ between the sexes. Aggression is generally related to reproduction. As males and females have different reproductive roles (see Topic N1), aggression serves different immediate ends for the two sexes. These sex-specific roles vary from species to species. In some, males maintain territories (and their resident female or females), which they protect by aggressive displays towards other males. In other species, males have to compete for rights to mate with the females. Such competition involves inter-male aggression. Females in a few species also have to compete with other females, for example, for nesting sites. More usually, however, female aggression is defensive, and is deployed in the defense of the young (see Topic N4).

Hormones and aggression

In laboratory animals, androgens have similar *activating* effects on offensive aggression as they do on sexual behavior (see Topic N2):

- male rats show far more offensive aggression than do females;
- castration of male animals reduces aggressiveness;
- inter-male aggression in rodents starts at the time of puberty, just when androgen secretion starts to rise;
- pre-pubertal mice can be made aggressive by injections of testosterone (McKinney and Desjardins, 1973).

They also have *organizing* effects:

- rats castrated in adult life respond rapidly with increased aggression to testosterone injection, while those castrated at birth do not respond.

Aggression in male rats is affected by social or situational factors, and this, also, interacts with sex hormones. A male rat housed with a female shows more aggression when confronted by a strange male rat than does a rat housed alone or with another male. This effect is increased if the animals are injected with testosterone (Albert et al., 1988). So the presence of a female sensitizes the male to the effects of testosterone. Similarly, castrated rats that were made to compete for food competed more strongly if they were administered testosterone. They subsequently attacked a docile, strange rat *only* if they had both testosterone and this previous competitive experience (Albert, Petrovic and Walsh, 1989).

Offensive aggression is rare in female rodents, but it does occur, and in different amounts in different individuals. This is also influenced by the activating and organizing effects of androgens. An activating effect is shown when adult females injected with testosterone show increased inter-female aggression (van de Poll et al., 1988). The organizing effect is shown in the individual differences both in natural response, and in response to adult testosterone injections. These result from differing exposure to androgens as fetuses. Rats have large litters, so a female fetus will be alongside different numbers of male fetuses (from none to 2; see *Fig. 1*). Those alongside two males have higher testosterone

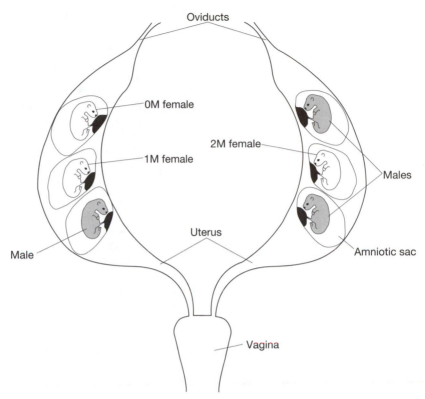

Fig. 1.   Exposure of female rodent pups to different amounts of testosterone, depending on
position in the uterus. Reprinted from van Saal (1983) in: Hormones and Aggressive
Behaviour (ed. B.B. Svare), with permission from Kluwer Academic/Plenum Publishers.

levels and respond more to adult testosterone injection than do those alongside
none. Their exposure to higher levels of androgens has a *masculinizing* effect
(vom Saal, 1983; see Topic N2). In many species most female offensive aggression is **maternal aggression** (see Topic N4). Female aggression in some species
has been shown to relate to hormonal changes in the estrus cycle (Mann, Konen
and Svare, 1984). Thus, female hamsters are aggressive except when in the
receptive phase of their estrus cycle. Ovariectomy renders them continuously
aggressive. Injection of both estrogen and progesterone reinstates their sexual
behavior and eliminates their aggressiveness (Floody and Pfaff, 1977).

**Neural
mechanisms of
aggression**

The two types of aggressive behavior have different central mechanisms (see
Gregg and Siegel, 2001, for a detailed account). The species-typical motor
pattern of *affective aggression* is programmed by neurons in the dorsal part of the
periaqueductal gray area (PAG) of the midbrain. These neurons excite neurons
in the brain stem and the spinal cord which produce the sympathetic and
behavioral aspects of rage. Lesions here stop the occurrence of affective aggression, but not predation. Stimulation produces affective aggressive behavior.
These PAG neurons receive inputs from the **medial hypothalamus**. Stimulation
of parts of the medial hypothalamus, particularly the anterior part, produces
affective aggression. However, lesions of the medial hypothalamus *do not* abolish offensive behavior, but change its probability of occurring in the presence of

particular sorts of stimuli, such as a male of the same species. So the medial hypothalamus normally plays a role in determining the occurrence of aggressive behavior, but it does not directly *initiate* it. Attack depends on the animal receiving particular stimulation, such as the presence of another animal, or sometimes tactile stimulation of limbs or face.

The anterior medial hypothalamus receives inputs from the **medial** and **cortical nuclei** of the **amygdala**. Stimulation of these locations facilitates (but does not by itself produce) affective aggression. Further modulation of aggressive behavior comes from an *inhibitory* pathway from the **basal nucleus** of the amygdala to the PAG, and an *excitatory* one from the **central nucleus**. The amygdala receives afferent inputs from many parts of the cerebral cortex, and sensory relays such as the parabrachial nucleus in the brain stem and the thalamus. It serves to integrate sensory and cognitive inputs, which then influence aggressive (and other) behavior (see Topic O2). The effect of cortical centers on affective aggression is shown by the removal of the cerebral cortex of cats, which results in uncontrolled **sham rage**.

Predatory attack is programmed by **brain stem** neurons controlled by input from the **lateral hypothalamus**. Mutual inhibitory connections between neurons in the lateral hypothalamus and the medial hypothalamus ensure that the two forms of aggression do not occur together.

Human aggression

In primates, environmental factors are the main influences on aggressive behavior. Whether monkeys are submissive, or dominant and aggressive, depends more on the level of aggression in the environment in which they are reared than on the individual's level of circulating androgens. Some female primates show variations in aggressiveness during the menstrual cycle, but the part of the cycle varies. In some it is greatest at the time of ovulation, at others immediately before menstruation. In humans too, aggression is related to activating effects of testosterone:

- aggression is far more frequent in men than in women. This difference is largest in pre-school-age children, moderate in 9–12-year-olds, and smaller in young adults (Hyde, 1986);
- the sex difference is largest for physical violence (but smaller when people believe that their behavior was unobserved, showing a strong social influence; Eagly and Steffen, 1986);
- human male aggression increases at puberty, along with increased testosterone secretion;
- castration of adult men usually decreases aggressiveness;
- higher levels of aggressiveness seem to relate to higher amounts of circulating testosterone androgens (Archer, 1994). However, the direction of causality is not clear.

It has been argued testosterone might be related to *dominance* rather than aggression, and the testosterone difference may be a *result* of changes in dominance, not a cause. In one study, people convicted of violent crimes, and dominant but non-aggressive convicts of *non*-violent crimes were equally high in testosterone, while non-dominant controls were lower (Ehrenkranz, Bliss and Sheard, 1974). In a series of studies, Mazur and Lamb (1980) showed similar relationships between dominance and testosterone. For example, testosterone was higher in men who had won a tennis match easily than in those who had won narrowly. Testosterone was elevated in people winning prizes as a result of their own efforts (students graduating), but not in those who won as no result of their own ability (lottery winners).

The inhibitory effect of the cerebral cortex on aggression in humans is shown by the fact that a number of disorders of the cortex are associated with increased aggressiveness. For example, a number of studies have linked violent crime in humans with lesions in the temporal lobe, including people with temporal lobe epilepsy (Monroe, 1978). The well-known fact that alcohol can increase aggressiveness has been attributed to it interfering with the executive functions of the prefrontal cortex, which normally control such behavior (Hoaken et al., 1998).

05 STRESS AND PSYCHOSOMATIC ILLNESS

Key Notes

The stress response	Selye's general adaptation syndrome describes the body's characteristic response to any stressor. It has three stages: the alarm reaction, the stage of resistance and the stage of exhaustion. Selye recognized two components of the stress reaction; the sympathetic–adrenal medulla axis and the anterior pituitary–adrenal cortex axis. Stress can also be viewed as resulting from allostatic load; accumulated wear and tear on the body, which disposes a person to illness.
Individual responses to stress	An individual must experience a stressor for the stress response to occur. The Social Readjustment Rating Scale has been used to predict health. But the impact of any life event depends on the individual's interpretation of the event, control over it, and adjustment to it. The individual modifies the environment, and modifies response to the environment by cognitive appraisal and coping responses.
Stress and illness	Some illnesses have been called psychosomatic illnesses, and were supposed to be the direct consequence of stress. But stimulus specificity suggests that different stressors have different disease-inducing characteristics. Individual specificity suggests that people respond to different stressors in specific ways. Coronary heart disease is associated with the Type A behavior pattern (TABP) of hostility, time urgency, achievement striving, competitiveness and impatience. The most important component of the TABP is thought to be hostility.
Psychoneuro-immunology	Modern psychosomatic medicine avoids these linear–causal explanations. Most illness is caused by multiple factors, including psychological events such as stress. Psychoneuroimmunology studies how stress influences illness through the immune system. One route is by the influence of hormones, including glucocorticoids and norepinephrine, on cells of the immune system.
Related topics	Peripheral factors in emotion (O1) Anxiety disorders (R3)

The stress response	Any situation that causes harm or threat to the person or the body is described as a **stressor**. The body's characteristic response to stressors is known as the **stress response**. Hans Selye (1936) noticed that people suffering from a wide variety of illnesses or **stressors** showed a common pattern of symptoms. Each stressor has specific effects on the body (e.g. cold causes shivering). In addition, any stressor produces a **general adaptation syndrome (GAS)**. The GAS has three stages (see *Fig. 1*):

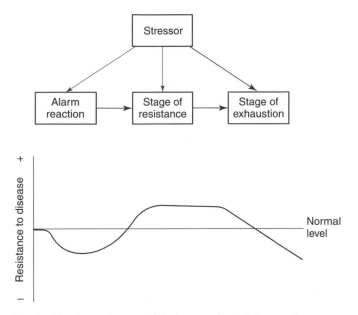

Fig. 1.   The three phases of Selye's general adaptation syndrome.

- the **alarm reaction**: the first exposure to the stressor produces characteristic changes, including enlargement of the **adrenal cortex** (see Topic E3), shrinkage of the **thymus** and **lymph nodes** (both parts of the immune system) and ulceration of the **stomach**. Resistance to infection etc. is decreased, and death may result if the stressor is strong enough;
- the **stage of resistance**: if the stressor is continued, the bodily changes of the alarm reaction disappear, and resistance increases above normal levels;
- the **stage of exhaustion**: further continuation of the stressor causes the physical changes of the alarm reaction to reappear, only now irreversibly. Resistance again falls, and the animal will die.

Selye recognized <u>two</u> components of the stress reaction, the **sympathetic–adrenal medulla** axis, and the **anterior pituitary–adrenal cortex** axis (see *Fig. 2*). In his own work, Selye concentrated on the second of these,

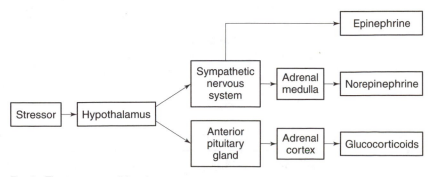

Fig. 2. The two axes of the stress response.

seeing the first as the initial reaction to emotional stimulation (compare Cannon's 'fight-or-flight' response, Topic O1). Stressors act on the **paraventricular nucleus** of the **hypothalamus**, which produces **corticotropin releasing factor (CRF)**. This passes to the anterior pituitary gland, causing it to secrete **adrenocorticotropic hormone (ACTH)**. This causes the secretion of **glucocorticoids**, including **cortisol**, from the adrenal cortex. The immediate effects of glucocorticoids include making energy available from stored fats and proteins (see Topic M1), and the promotion of tissue repair. CRF is also produced in other brain centers, notably in the limbic system, where it acts as a neurotransmitter serving emotional responses.

The longer-term effects of stress have recently been viewed in terms of **allostatic load** (see McEwen and Seeman, 2003). **Allostasis** is the maintenance of stability in the body through physiological change. It is a longer-term equivalent of homeostasis (see Topic L1). The components of the stress response serve this maintenance of stability. However, prolonged or repeated stress leads to allostatic load, which is accumulated wear and tear on the body which disposes a person to illness.

Individual responses to stress

The stress response occurs not only to physical stressors, but also to psychological ones, such as perceived danger, or illness of a close relative. Furthermore, even physical injury by itself is not enough to produce the stress response. For example, major surgery under general anesthesia does not produce a stress response, whereas the same operation *without* anesthesia would very likely do so. The person needs to *experience* the stressor for the stress response to occur.

This led in the 1960s to greater attention being paid to the individual in relation to sources of stress, response to stressors, and ways of alleviating stress. Holmes and Rahe (1967) produced the **Social Readjustment Rating Scale**. This is a list of 43 **life events**, each with an associated score of life change units; a rating of how stressful it is. The items range from minor violations of the law (11 units) to death of a spouse (100 units). Using this, they showed that the total impact of recent life events predicted subsequent health. However, later results were less clear, mainly because the scale assumes each event has the same impact on everybody. On the contrary, the impact of any life event depends on:

- **interpretation**: each individual interprets (or appraises) an event in relation to their own circumstances, needs, motivation, etc.;
- **control**: individuals differ in how much control they have over an event. Greater control results in less impact;
- **adjustment**: partly independent of interpretation and control, people require different amounts of adjustment to an event.

This leads to a **transactional** view of stress. In this view, stress is an interplay between the person and the environment. The individual is not a passive respondent to environmental stressors, but modifies the environment, and modifies response to the environment by **cognitive appraisal**, and various **coping responses**.

Stress and illness

Selye called certain illnesses **diseases of adaptation**. These diseases, including gastric ulcers, asthma, arthritis, hypertension and eczema, have also been called **psychosomatic illnesses**. The supposition was that certain illnesses are the direct consequence of exposure to stressors. This gives rise to a problem of specificity. If the GAS is a nonspecific response to a nonspecific stressor, why do people develop different psychosomatic illnesses? One answer is to deny the nonspecificity of the stress response, and argue for **stimulus specificity**; that different stressors do have different disease-inducing characteristics. Alternatively, some theorists have argued for different types of **individual specificity**; that people respond to different stressors in ways that are specific to the individual. Thus, for example, some will respond to a stressor with a cardiovascular response, and develop cardiovascular disease. Others might respond to the same stressor with a gastric response and end up with ulcers. A different view related specific illnesses to the suppression of expression of specific emotions, for example asthma with suppressed anger. Alexander (1950), from a Freudian perspective, proposed, for example, that asthma resulted from unresolved dependence on the mother, and migraine from repressed hostile impulses.

A more recent example of this sort of specificity view is the linking of **coronary heart disease** with the **Type A behavior pattern** (**TABP**). Friedman and Rosenman (1959) showed that people who had suffered heart disease were characterized by a personality type they called the Type A behavior pattern. The characteristics of this are:

- hostility;
- time urgency;
- achievement striving;
- competitiveness;
- impatience.

In a prospective (that is predictive) study, they showed that persons with TABP are more than twice as likely to subsequently suffer heart disease than are others (Rosenman et al., 1964).

Psychoneuro-immunology

Modern psychosomatic medicine avoids the simplistic **linear–causal** explanations of classical psychosomatic medicine. The classic 'psychosomatic illnesses', as well as increased vulnerability to other illness, are part of a multifactorial stress response (or allostatic load), itself caused by multiple factors. An individual's response to stressors is regulated by genetic, learned and socioeconomic factors. A good example of the shift in emphasis is the case of gastric ulcers. The classical view that these are a direct result of stress dates back to Selye's observations noted earlier. Stress produces excessive acid production and reduces gastric motility and blood flow, resulting directly in ulceration. However, in the 1980s it was discovered that almost all cases of gastric ulcers not caused by taking drugs like aspirin were associated with a bacterium living in the gut: *Helicobacter pylori*. This led some physicians to claim that this was the simple cause of ulceration. However, it was soon noted that 75% of people without ulcers also have *H. pylori* in their gut. Furthermore, psychological treatment to reduce stress reduces ulceration without changing the infection with *H. pylori*. The most likely explanation is that ulceration results from a combination of the effects of bacteria and stress (Overmier and Murison, 1997).

The ways in which stress acts to influence the onset and course of illness are studied in the field of **psychoneuroimmunology**. The immune system consists of organs and cells specialized to protect the body against invasion by foreign organisms. The cells involved fall into two broad classes. **Phagocytes** are cells that nonspecifically engulf and destroy foreign bodies. **Lymphocytes**, which carry out specific attacks on foreign cells, are of two types. **B cells** react to foreign cells by producing specific **antibodies**, which attach to **antigens** on the foreign cell and thereby inactivate or destroy it. **T cells** (or killer cells) are attracted to phagocytes that have engulfed a foreign cell, and change so as to reproduce in a form that will attack the specific type of foreign cell.

Psychoneuroimmunology studies the mechanisms by which stress (and other psychological processes) influence the immune system. There is ample evidence that stress suppresses immune function. One route through which this could operate is by the influence of hormones on T and B cells. Both types of cell have receptors for glucocorticoids and norepinephrine (McEwen et al., 1997).

P1 THE NATURE OF LEARNING AND MEMORY

Key Notes

Defining learning	Defining learning is not easy. We need to consider both when a piece of learning actually takes place and where in the brain this learning occurs.
Types of learning	For our purposes, learning can be non-associative or associative. Non-associative learning occurs through habituation or sensitization. Associative learning includes classical and operant conditioning.
Types of memory	There are many ways in which memory can be sub-classified. Distinctions between short-term and long-term memory, and between working memory and reference memory, have been useful in the study of the neural correlates of learning.
The engram	An engram is an actual memory trace and the idea that one could pinpoint exact neurons in which learning had taken place became known as the search for the engram. Critics of this view see memories as distributed patterns of activity across large regions of the brain.
Related topic	Classical and instrumental conditioning (P2)

Defining learning

Physiological psychologists need to be able to define learning in order to investigate the processes that form the basis of learning. However, finding a useful definition of learning is not as easy as it might seem. Try to recall *exactly* when it was that you learnt anything. The physiological psychologist might need to know *when* the learning occurred to a few milliseconds if the learning is to be related to specific brain events. If the question of when proves too difficult then the question of *where* becomes unanswerable. By the same token, if you have no idea of where in the brain to look then you have little chance of observing the crucial learning event. Furthermore, an observed brain event can only be a candidate for learning if it can be distinguished from a perceptual event or a motor event, and so on. Of course, we are assuming here that learning is characterized by a discrete event, when it might equally be a sequence of events, and that sequence might occur in various parts of the brain.

Some of the attempts at solutions to these problems are the subject of this Section. In this Topic a number of the important concepts are explained. In Topic P2 some of the paradigms used are examined. Topics 3 and 4 give an overview of some of the most interesting research into where and how learning occurs. Topic P5 considers research into amnesia.

Types of learning Learning can be simply divided into associative and non-associative forms. **Non-associative learning** is the simplest learning that an organism can engage in. It includes habituation and sensitization. **Habituation** is when the response to a stimulus lessens as a result of repeated exposure. An example might be how church bells can wake you up on a Sunday morning but very shortly afterwards you no longer notice them. However, habituation is distinguishable from adaptation or fatigue, both of which would also cause a lessened response.

Sensitization is conceptually the reverse of habituation, although the neuronal responses that are responsible for it are very different. If you were given a mild electric shock to your arm you would doubtless react to it. If the same level of shock was repeated again and again you would most likely start reacting with an increased response. This increased response is the result of sensitization. Both habituation and sensitization are important forms of learning for any organism. Habituation allows an organism to ignore non-threatening stimuli and sensitization serves to remind an organism of an unpleasant previous encounter.

Associative learning is where learning takes place as a result of a new association between two or more events. **Classical conditioning** occurs as a result of the pairing of a conditioned stimulus (CS) with an unconditioned stimulus (UCS). The CS does not initially evoke a response but the UCS does. After a number of pairings, the CS comes to evoke the same response as the UCS. A study involving this kind of learning is described in detail later in the section. **Instrumental conditioning** is learning by consequence. Simply put, if a behavior is performed and the consequence of that behavior is pleasant, then the behavior is likely to be repeated. If the consequence is unpleasant, it is likely not to be repeated so readily in the future. This form of learning differs from the others described so far, as the behavior displayed is usually more complex. Hence, it is more difficult to identify learning using instrumental conditioning. Nevertheless, instrumental conditioning can be used to investigate the consequences of lesions on learning and memory.

Types of memory As well as being able to distinguish different types of learning, it is useful to distinguish between the different types of memory that can be investigated. Distinctions that have emerged from cognitive psychology, such as short-term versus long-term memory, or working memory versus reference memory, have been useful in the study of neural correlates. For example, lesions of certain brain areas might selectively disrupt one type of memory whilst leaving other types intact. In Topic P2 we will see that the work of Olton points to the hippocampus being necessary for working memory whilst studies by Morris suggest that this region is needed for reference memory.

The engram The term **engram** refers to the idea that there is an identifiable neuronal event that is associated with the storage of a memory — the **memory trace**. This idea stems from the notion that individual neurons learn. Hence the engram is represented by a change in a neuron's output characteristics as a result of having learned something. The 'search for the engram' has been hotly debated, with some critics suggesting that memories are coded via patterns of activity in groups of neurons rather than by a specific change in one neuron. Nevertheless, the broader notion of the engram as an identifiable piece of memory is still useful within certain limits, as we will see when we look at the work of Thompson and his colleagues (Topic P4).

P2 CLASSICAL AND INSTRUMENTAL CONDITIONING

Key Notes

The classical conditioning paradigm	Classical conditioning is a simple form of associative learning but provides the high degree of control over the learning situation necessary to establish its neural correlates. A conditioned stimulus (CS) is paired with an unconditioned stimulus (UCS), and comes to elicit the response originally elicited by the UCS. The critical changes in behavioral response to the CS and UCS are exactly matched by changes in hippocampal and cerebellar activity.
The instrumental conditioning paradigm	The instrumental conditioning paradigm affords the opportunity to examine the learning of voluntary behaviors. Reinforcement and punishment are strong determinants of whether or not behavior will be modified as a result of learning.
Learning about reward and punishment	Animals and people will respond at very high rates in order to stimulate certain areas of the brain such as the median forebrain bundle and the periventricular tract. Dopaminergic systems in these areas are the basis of reinforcement, crucial to learning based on punishment and reward.
Working memory and reference memory	Working memory is memory needed only in the here and now, whereas reference memory is needed time and again. Conflicting research in the 1980s on the role of the hippocampus in working and reference memory was resolved by showing that the hippocampus is necessary for remembering spatial information.
Related topics	The nature of learning and memory (P1) Brain structures involved in memory (P3)

The classical conditioning paradigm

Classical conditioning represents the simplest form of associative learning. First discovered by Ivan Pavlov (1927), it is the form of learning that occurs when a stimulus that is originally neutral comes to elicit the same response as a non-neutral stimulus because of a learned association between the two stimuli. It has become a useful learning paradigm because it is relatively easy to establish, it is robust and reliable across many species, and it can be used to cause learning of a discrete motor response. This latter feature is important if we are to try to identify the neural correlates of the point of learning as discussed in Topic P1.

The classical conditioning paradigm concerns the pairing of two stimuli, the **conditioned stimulus (CS)** and the **unconditioned stimulus (UCS)**. The CS is originally a neutral stimulus that elicits no particular response. The UCS is a stimulus that always elicits a response, the **unconditioned response (UCR)**.

When the two are paired together in the right way, the CS comes to elicit the same response as the UCS. The response to the CS is called a **conditioned response (CR)**. This conditioning works best when the CS starts just before the UCS. This type of conditioning is also easier to achieve for involuntary behaviors, such as reflexes, than for voluntary behaviors.

A classic example of this conditioning that is relevant to the topic of learning and memory is the **rabbit nictitating membrane response**. Here the CS is a light or an audible tone and the UCS is a puff of air to the cornea. The nictitating membrane is a third eyelid that closes across the eye from the nose laterally. By holding the other two eyelids open and delivering a puff of air to the cornea, a reflex is initiated that causes the eyeball to retract and the nictitating membrane to extend across the eye. Trials are delivered in groups of nine, with the first eight consisting of CS and UCS delivery and the ninth trial being a CS alone trial. The purpose of this ninth trial is to allow the easy measurement of any CR that has occurred. With the stimuli set up as shown in *Fig. 1*, a strong CR develops over time.

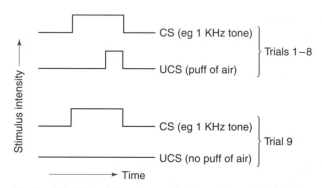

Fig. 1.   CS–UCS relations in conditioning of the rabbit nictitating membrane response.

This paradigm has been used by Thompson and co-workers to demonstrate that the **hippocampus** is involved in this type of learning. However, the hippocampus is not *critical* to the learning or retention of this learned response. Instead, that role goes to the **cerebellum**. It is probable that the hippocampus monitors this learning in case the level of learning required goes beyond the capability of the cerebellum. If the hippocampus were not monitoring the situation, it would be hard for it to take over once the cerebellum could no longer cope.

Given that the cerebellum is a phylogenetically older part of the brain, we can see a pattern emerging. Earlier forms of learning were simple associative forms of learning. As the capacity of the brain for the manipulation of information developed, phylogenetically newer parts of the brain were needed to take on the learning and memory requirements.

The instrumental conditioning paradigm

This type of conditioning is concerned with **rewards** and **punishments** that are a consequence of voluntary behavior. First characterized by Skinner (1938), the paradigm reflects the fact that when an organism acts it does so because of the prevailing conditions (the antecedents). There are also consequences to the behavior that can be pleasurable, neutral or unpleasurable. To describe this, Skinner coined the term **ABC** (antecedents, behavior, consequences).

Furthermore, if the consequence is pleasurable, the behavior is more likely to be repeated and if it is unfavorable the behavior is less likely to be repeated. In other words, the behavior may or may not be *reinforced*.

Reinforcement can be positive, in the form of rewards, or negative, in the form of witholding a reward. Punishment is another form of negative reward but is different from negative reinforcement as *Table 1* demonstrates. Interestingly, punishment does not produce as strong a learning effect as negative reinforcement.

Table 1. Types of reward and their consequences for behavior

Reward type	Explanation	Outcome
Positive reinforcement	The behavior is rewarded. For example, a child is given sweets for good behavior.	Behavior is more likely to be repeated in future.
Negative reinforcement	A reward is withdrawn. For example, a child is not permitted to attend a party because of poor behavior.	Behavior is less likely to be repeated in future.
Punishment	The behavior is punished. For example, a child is smacked for exhibiting poor behavior.	Behavior is less likely to be repeated in future.

Learning about reward and punishment

Early work by Olds and Milner (1954) showed that rats will press a lever in order to receive an electrical stimulus to **limbic system** areas such as the **septum** (a phenomenon called **self-stimulation**). Indeed, they find the stimulation so rewarding, they will press at rates of 2000 presses per hour. Monkeys will even press at a rate of 8000 per hour (Olds, 1962). Clearly, these animals find the stimulation very rewarding. Later work showed similar results in humans, who described the self-stimulation as nonspecifically pleasurable.

More recent research has found that an area of the brain called the **median forebrain bundle** is a reliable area for self-stimulation. An increase in **dopamine** in this area is involved in the mechanism by which self-stimulation is enhanced. The median forebrain bundle reinforces the connection between the system that detects the pleasurable stimulus (e.g. seeing some food) and the motor system that carries out the desired behavior (e.g. eating). In other words, this region of the brain carries the mechanism of reinforcement.

Working and reference memory

Another area of research that uses instrumental conditioning to investigate memory has linked biological psychology to concepts used in cognitive psychology. **Working memory** refers to knowledge that is required to be remembered in the here and now but that can be forgotten later (a classic example is remembering a telephone number long enough for it to be dialed). **Reference memory** is the term used to describe long-term knowledge that is potentially needed time and again.

Research investigating the role of the hippocampus sparked off an interesting debate between David Olton and Richard Morris. Morris et al. (1982) showed that rats could learn to swim through murky water in order to find a podium located just beneath the surface. Morris claimed that the rats used external cues to orientate themselves in space and that removing the hippocampus disabled the rats from finding the podium. As this spatial information is part of reference memory, Morris claimed that removing the hippocampus disrupted reference memory.

Olton (1983) was also investigating reference memory, but together with working memory. He devised an eight-arm radial maze (*Fig. 2*) in which the same four arms were baited with food each time. The rat had to learn to visit only the baited arms (reference memory) and only to visit each arm once on any given trial (working memory). Olton discovered that removing the hippocampus disrupted the working memory component (they now entered baited arms again even though they had got the food from there) but left the reference memory component intact (they still never entered the unbaited arms).

Although these two sets of results appear contradictory, there is a commonality in that both tasks require the remembering of spatial information. Indeed, researchers (e.g. O'Keefe and Nadel, 1978) have found strong relationships between the firing of hippocampal neurons and the remembering of place information.

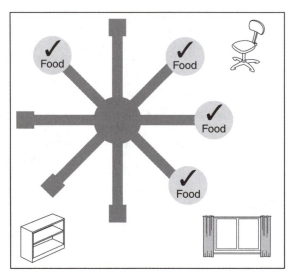

Fig. 2.   Olton's eight-arm radial maze.

P3 BRAIN STRUCTURES INVOLVED IN MEMORY

Key Notes

Spinal memory	Spinal memory occurs through the alteration of firing patterns of neurons in the spinal cord. Learned changes in sensitivity to pain can be caused by alterations at the level of the spinal cord.
The cerebellum	The cerebellum is a likely candidate for the acquisition and storage of motor memories. Areas of particular interest are the deep cerebellar nuclei, although the cerebellar cortex is also believed to be involved.
Diencephalic structures	A number of diencephalic structures have been implicated in having a role to play in the formation and maintenance of memories. In particular, the thalamus and the mammillary bodies have been strongly implicated.
The hippocampus	The hippocampus seems to be centrally involved in many aspects of memory. Its internal structure is one that enables researchers to accurately pinpoint the location of recording electrodes.
The limbic system	Many limbic structures have been linked to amnesia and hence to the maintenance and/or recall of memories.
Related topic	The anatomy of the central nervous system (C1)

Spinal memory

There is evidence that simple learning can take place even at the level of the spinal cord. Such learning involves alteration of the firing patterns of neurons involved in reflexes and has been shown in rats whose spinal cords have been transected. For example, Lopez-Garcia (2003) has shown that spinal memory is involved in mediating pain responses. There is also evidence that paraplegics can be re-taught some useful movement if treatment is started within 72 hours of the injury. Any motor learning in such patients must be happening in the spinal cord.

The cerebellum

The cerebellum has already been referred to in the discussion about the rabbit nictitating membrane response (see Topic P2). As a phylogenetically older part of the brain and a motor control center, the cerebellum is a likely candidate for a role in memory. After all, many of our most basic motor memories are under automatic control (e.g. walking), as are our more complex but frequently expressed motor sequences (e.g. making a cup of coffee).

There has been some debate about which part or parts of the cerebellum may be involved in memory formation and retention. Research by Thompson's group (e.g. Lavond et al., 1987) favors the deep cerebellar nuclei whereas work by Yeo's group (e.g. Yeo, 1984) favors the cerebellar cortex. Either way, it is certain that the cerebellum plays a major role in many motor memories.

Diencephalic structures

We will see the importance of the diencephalon when we come to review some neuropsychological disorders in Topic P5. For now, we can concentrate on those diencephalic structures that have been implicated in memory. Of the two major structures in the diencephalon, it is likely that parts of the thalamus are involved in the diencephalic memory system and unlikely that the hypothalamus is involved. However, a number of other structures have been implicated, such as the mammillary body, although there is some debate about how critical to memory these structures are.

The hippocampus

We have already mentioned the hippocampus a number of times in this chapter and the evidence for its involvement is very strong. It is worth noting here that the hippocampus is very easy to record from as it is a laminar structure. This means that it has a uniform cross-section. Cut the hippocampus anywhere perpendicular to its length and the cut end looks the same (as you would find doing the same to a cucumber). It may be no chance event that the cerebellum, too, is laminar. Whether this is a coincidence or reflects some electrical properties of these structures is a matter for speculation.

We will consider another feature of the hippocampus, that of **long-term potentiation**, in Topic P4. This phenomenon has provided our best understanding to date of the possible mechanisms by which learning takes place. Again, it is, in part, the electrical properties of the hippocampus that allow such an in-depth analysis of the biochemistry involved.

The hippocampus is also a structure that is damaged in some amnesics. The patient known as HM, for example, was profoundly amnesic following bilateral damage to the hippocampus. This will be considered in more detail in Topic P5.

The limbic system

Some of the structures that we have already mentioned are part of the **limbic system** (see Topic O2). Many of the structures that make up the limbic system have been proposed as having some involvement in the learning process. *Fig. 1* shows the limbic structures that have been implicated. Most of these have been suggested as a result of neuropsychological patients who have expressed a degree of amnesia and who have damage to one or more of these brain areas.

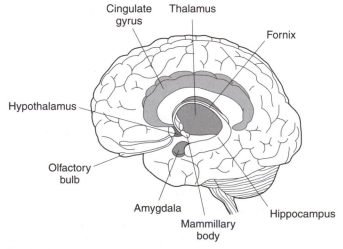

Fig. 1.   Structures in the limbic system that are involved in memory.

P4 THE NEURAL BASIS OF LEARNING AND MEMORY

Key Notes

Early thinking	Researchers like Lashley, Penfield and Hebb were responsible for the first attempts to systematically study the effects of learning on the central nervous system. They believed memory was located in the cortex, either localized or distributed.
Invertebrate learning	Studies by Kandel and others of simple neural systems point to learning taking place through alteration of synaptic functioning. In habituation, the amount of neurotransmitter released by the stimulus diminishes. In sensitization, depolarization is prolonged, increasing the amount of neurotransmitter released.
Long-term potentiation	Long-term potentiation is a change in neuronal firing rate, maintained over a long period of time. It is acquired following activation of an N-methyl-D-aspartate (NMDA) receptor causing an influx of Ca^{2+} ions. The maintenance mechanism involves an increase in the amount of neurotransmitter released by the presynaptic neuron.
The neural basis of the engram	In studies of learning in mammals using electrophysiology and lesion studies, researchers have found that changes in the hippocampus and the cerebellum track learning. Lesions in the hippocampus do not affect the conditioned response, but those in the cerebellum do. Thus suggests the cerebellum as the site of the engram.
Cognitive maps	Neuroethologists, such as O'Keefe and Nadel, have discovered that the hippocampus might be involved in the establishment and/or maintenance of cognitive maps.
Related topics	The nature of learning and memory (P1) Brain structures involved in memory (P3)

Early thinking

It is worth taking a brief look at the early research by Karl Lashley, Wilder Penfield and Donald Hebb, as many of the current debates stem from these beginnings. Lashley (e.g. 1929) was probably the first researcher to study learning in the brain systematically. He believed that memory was a feature of the cortex and set about trying to discover where this function resided. He was unable to locate the seat of memory (the engram; see Topic P1) even though he believed that such a place must exist.

Penfield and Perot (1963) also believed that memory was cortical but their conclusion came as a result of surgery on humans. In order to pinpoint an area to be lesioned, an area of cortex is exposed during surgery and stimulated. The patient's response enables the surgeon to operate on the correct area. During this procedure, Penfield noted that if the stimulation was in the association cortex, memories could be elicited. He concluded that these memories must reside in this area of the brain.

In contrast to the other two, E. Roy John's idea of memory was quite different. John (1972) believed that memories were distributed throughout the cortex and that any one memory did not reside in any one place. Hebb (1949) also believed in a distributed memory system and his ideas are the foundation of one form of modern-day connectionist architecture. This principle of **parallel distributed processing** is in direct opposition to the idea of the engram. The modern picture is far from clear and it is possible that both kinds of system exist. Nevertheless, these early beginnings sparked the massive amount of research that has been undertaken over the last 50 years.

Invertebrate learning

For some researchers, the idea of trying to understand the complexities of mammalian memory is too far in the future. These researchers have opted to try to explore the biochemical underpinnings of memory in invertebrates. Kandel and co-workers (Kandel and Spencer, 1968; Kandel et al., 2001) have been investigating simple learning in the sea snail called Aplysia. Aplysia has a simple neuronal network that has been fully described. Learning in such (relatively) simple systems is restricted to habituation, sensitization and simple forms of associative learning (see Topic P1). Kandel has discovered the biochemical mechanisms responsible for these types of learning. Such discoveries are extremely important because, given the principles of evolution, we might expect to find that mammalian learning shows many similarities. After all, nature usually builds on success rather than completely reinventing the wheel.

The basic findings of Kandel point to learning taking place through an alteration of synaptic functioning (see Topic B3). In the case of *habituation* (Topic P1), the postsynaptic mechanism is unchanged but the amount of neurotransmitter released by the stimulus diminishes over successive trials. This is accomplished by reducing the amount of Ca^{2+} ions that enters the terminal bouton when an action potential travels along the sensory neuron. For *sensitization*, the mechanism is slightly different and involves a facilitatory neuron whose action leads to a reduction of the K^+ efflux from the neuron. This prolongs the depolarization phase of the action potential and thereby increases the amount of neurotransmitter released.

Long-term potentiation

Long-term potentiation (LTP) was first discovered by Bliss and Lømø (1973). They found that if the **perforant pathway** in the **entorhinal cortex** of the brain was repeatedly stimulated at a fast rate (a tetanus) then the responses to a single stimulus in the receiving **dentate gyrus** of the hippocampus would be potentiated. Furthermore, using a hippocampal slice preparation, this potentiated response could be seen to last for several days or even weeks. Further research unearthed the mechanism by which this potentiated response occurs.

The mechanisms for acquisition and maintenance of LTP are different. The acquisition mechanism is postsynaptic and involves a large influx of Ca^{2+} ions via activation of an NMDA receptor. This receptor is a particular type of glutamate receptor that is normally inactive because it is blocked by magnesium ions. However, the burst of stimuli delivered through the tetanus removes the magnesium block and thereby activates the receptor. The Ca^{2+} ions then set off a chain of events postsynaptically, leading to a persistent increase in transmission across the synapse. The maintenance mechanism is presynaptic and involves an increase in the amount of neurotransmitter released by the presynaptic neuron. There is much debate about the means by which the postsynaptic acquisition event gets translated into a presynaptic maintenance event. The debate arises because the acquisition mechanism is postsynaptic and the maintenance mechanism is presynaptic. The question, then, is how does the postsynaptic cell communicate back to the presynaptic cell? One suggestion is that a chemical messenger travels back from the postsynaptic cell to the presynaptic cell. It is likely that one such messenger is nitric oxide but the pre- and postsynaptic mechanisms are so complex that it is quite possible that more than one messenger is involved.

The neural basis of the engram

The work of Thompson and his colleagues was referred to in Topic P2. Using the **rabbit nictitating membrane** paradigm, Thompson's group has been able to show that hippocampal multiple unit activity (activity recorded from several neurons) looks identical to the pattern of behavioral responding (i.e. the movement of the nictitating membrane) during both acquisition and retention. *Fig. 1* shows the hippocampal activity and behavioral response associated with paired training trials before and after learning has taken place. You can clearly see how the two traces change in an identical way as learning has taken place. Unfortunately, Thompson found that lesioning the hippocampus had no effect on either the acquisition or retention of the behavior. However, similar findings were obtained from recordings of the deep cerebellar nuclei. Here, lesions differentially knocked out the conditioned response whilst leaving the unconditioned response intact. This was true for both acquisition and retention. Thompson concluded that the engram for the rabbit nictitating response resides within the deep cerebellar nuclei. Whilst there has been some dispute with Yeo over the precise critical location in the cerebellum, it would appear that the cerebellum is a necessary structure for this form of learning. If we ask what the hippocampus is doing in tracking this task in the first place, we might surmise

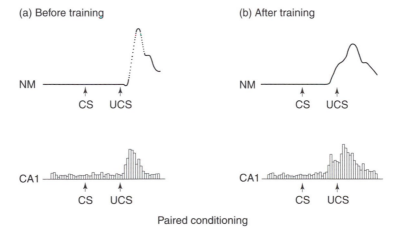

NM = Nictitating membrane (behavioral response)
CA1 = Region CA1 of the hippocampus (neuronal response)

Fig. 1. Hippocampal and behavioral responses for the classical conditioning of the nictitating membrane response.

that the hippocampus is involved in case the complexity of learning required becomes more than the cerebellum can cope with. This would make sense in the context of evolution, with a phylogenetically newer part of the brain adding functionality rather than replacing an older system.

Cognitive maps   Not all researchers have taken such a reductionist approach to the study of memory. O'Keefe and Nadel are neuroethologists and have asked the more holistic question: what does the hippocampus do? One of the things that they found (O'Keefe and Nadel, 1978) was that the hippocampus is involved in spatial memory. They discovered cells in the hippocampus that respond maximally when the organism is in a particular location in space. They called these cells **place cells**. O'Keefe and Nadel argue that the hippocampus is where we generate and/or store our cognitive map of our environment.

P5 AMNESIA

Key Notes

Types of amnesia	The neuropsychological disorder of amnesia has been sub-classified along a number of different dimensions. It can be retrograde or anterograde (time), organic or psychogenic (origin), episodic or semantic (type of loss).
Causes of amnesia	Organic amnesia can have many causes, including viruses, strokes, brain injury and tumors. The pattern of amnesia depends on the structures affected. Inorganic amnesia follows psychological trauma.
Anatomy of amnesia	The type of amnesia that a person suffers from is partly dependent on where the damage has occurred. Two distinctive regions are the diencephalon and the temporal lobes. In addition, whether or not there is frontal damage can be critical in determining how the amnesia manifests itself.
Wernicke–Korsakoff syndrome	This is the most well studied form of amnesia. Described at around the same time by Karl Wernicke and Sergei Korsakoff, it is a form of diencephalic amnesia, often resulting from alcohol abuse. A frequent sign is confabulation.
Temporal lobe amnesia	As the name suggests, this is amnesia caused by damage to the temporal lobes. The pattern of deficit is different to that which accompanies Wernicke–Korsakoff syndrome.
Post-encephalitic amnesia	Although this is also a form of temporal lobe amnesia, its origin is a form of the Herpes simplex virus and the pattern of deficit is somewhat different to that of the temporal lobe amnesia described earlier.
Models of amnesia	Amnesia can result from either a failure to encode, a failure to store, or a failure to retrieve. A great deal of research has attempted to determine the appropriate model for Wernicke–Korsakoff amnesia.

Types of amnesia

Amnesia is not a unitary phenomenon. A particular patient will have a pattern of memory problems that will reflect the precise area of damage. In order to distinguish between different types of amnesic problem, there are a few terms that are used to describe the various types of deficit seen. An amnesic problem can be organic or inorganic (psychogenic). An **organic** disorder is one where there is some clearly discernible brain damage whereas an **inorganic** disorder is usually psychological in origin. An amnesia can also be either retrograde or anterograde. **Retrograde amnesia** refers to an inability to remember something that occurred prior to the injury whereas **anterograde amnesia** refers to the inability to remember new things after the injury. Another distinction often seen between amnesics is deficits in episodic versus semantic memory. **Episodic memory** is the ability to remember personal events such as what you had for breakfast. **Semantic memory** is the ability to recall knowledge such as the capital of France.

Causes of amnesia

Organic amnesia is always the result of brain damage. However, such damage can be caused by a number of different types of event. These include viruses, strokes, head injury, tumors, anoxia (lack of oxygen) and so on. A virus such as a strain of the herpes simplex virus can attack the brain and cause damage leading to an amnesic syndrome. Likewise, a stroke that constricts the posterior cerebral artery can cause an ischemia to the hippocampus, resulting in amnesia. Whatever the cause of the amnesia, it will almost certainly be the case that the resulting lesion will have damaged more than one structure (see Topic A2). The pattern of damage typical from these various causes has led to a number of separately identifiable amnesic syndromes. Equally, though, the varied nature of each individual's pattern of damage has produced a great deal of controversy over whether or not there are critical anatomic structures that must be damaged for any amnesia to occur.

Inorganic (psychogenic) amnesia is characterized by a lack of obvious brain damage and is usually associated with a traumatic event. For example, hysterical amnesia is often associated with an event like death of a relative or with combat duty. However, an extreme form, known as a fugue state, can involve a complete loss of identity. During the fugue, the person remembers nothing of who they are or where they are from. The fugue can last from hours to years and when the fugue state dissipates little can be remembered of the fugue period. Other sources of psychogenic amnesia can be accidents. The term 'compensation neurosis' has been coined to describe people who suffer a head injury and then claim a level of amnesia that is beyond what one might expect, given the injury sustained. Often these are people awaiting the settlement of an insurance claim, hence the label.

Anatomy of amnesia

Amnesic syndromes appear to be split into two broad types on the basis of their gross anatomical origins. One type of amnesia originates from damage to diencephalic structures (see Topic C1). Two structures have been strongly implicated as critical centers of damage in **diencephalic amnesia**. These are the **dorso-medial nucleus of the thalamus (DMTN)** and the **mammillary bodies** (not strictly in the diencephalon, but part of the limbic system; see Topic P3). The data do not point clearly to whether one or both of these structures must be damaged in order for diencephalic amnesia to occur.

Another type of amnesia originates from damage to the **temporal lobes** of the cerebral cortex. Again, more than one structure has been cited as the critical region for damage. The main suggestions are the **hippocampus** and the **amygdala**. As for diencephalic amnesia, the critical region of damage is not clear from the data.

Even though there is much debate about sites for critical lesions, there is little doubt that the limbic system is a major player. Furthermore, it seems that these two separate areas of damage give rise to different kinds of amnesic syndrome.

Wernicke–Korsakoff syndrome

This form of amnesia was discovered at about the same time by Carl Wernicke and Sergei Korsakoff. It is a diencephalic amnesia and is typified by poor short-term memory, and an anterograde rather than retrograde deficit in long-term memory. However, some cases of retrograde amnesia are reported in the literature (Stuss et al., 1988).

Wernicke–Korsakoff syndrome (WKS) is often the result of long-term alcohol abuse. Some have suggested that the syndrome is the result of a poor diet, which alcoholics often have. More likely, though, is that it is a combination of the toxic effects of alcohol, together with the poor diet, that results in the dien-cephalic damage. A typical feature of these patients is **confabulation**; describing a fictitious event instead of recalling something that actually happened. This is especially likely when the deficit is first noticed. The probable cause of this is a denial that anything is wrong. However, the confabulations can be wildly untrue and far-fetched, suggesting that this cannot be the whole explanation. Also often accompanying WKS is a severely impaired performance on sequence-switching tasks like the Wisconsin Card Sorting Task. This is the result of frontal lobe damage that invariably accompanies the damage to the diencephalon.

Temporal lobe amnesia

As the name suggests, this form of amnesia derives from bilateral damage to the temporal lobes. A famous case of this amnesia was Scoville and Milner's (1957) patient, HM, who was given a bilateral temporal lobectomy to relieve the symptoms of epilepsy. Unlike diencephalic patients, most temporal lobe amnesics have normal short-term memory. They usually have a severe antero-grade amnesia and a degree of retrograde amnesia that varies from patient to patient. Temporal lobe amnesics do not suffer from confabulation or from any frontal symptoms.

Post-encephalitic amnesia

In very rare cases, infection with the Herpes simplex virus can result in Herpes simplex encephalitis (HSE). Primary damage occurs in the **medial** and **lateral temporal cortex**, causing an immediate post-encephalitic amnesia. As with other causes of temporal lobe amnesia, short-term memory remains intact. Typical cases exhibit very severe anterograde and retrograde amnesias. Their memory can last for only a few minutes and some patients can become confused and disoriented about space and time. There are some signs of frontal damage but they are much less severe than with WKS patients.

Models of amnesia

One thing is common among all amnesics and that is that their memory is poor. However, there has been much debate about the reasons for this poor memory. Some have suggested that amnesics fail to properly encode material, whilst others claim that amnesics have poor retrieval strategies. The third possibility is that amnesics fail to store memories but that they encode the material properly and could retrieve it were it there.

Most of the research in this area has been carried out on WKS patients. The most favored suggestion is the encoding deficit hypothesis. This fits data that have shown that if amnesics are given longer than controls to encode material their performance can be just as good at retrieval. Furthermore, it appears that WKS amnesics are particularly poor at encoding context information (the context deficit hypothesis). Information like when, where, and so on fail to become encoded along with the 'to-be-remembered' item itself. This means that their intact retrieval mechanisms have fewer cues with which to effect an accurate retrieval.

Q1 LANGUAGE AND ITS DISORDERS

Key Notes

Language	Aphasia is an impairment of language production or comprehension. Agraphia is the inability to produce written language, and dysgraphia is an impairment of written language. Dyslexia refers to disorders of reading.
Broca's aphasia	Broca's aphasia refers to a range of language difficulties resulting from damage to the left frontal cortex. Some patients cannot speak at all. Others can speak with effort, often using wrong words and simple grammatical constructions, and tending to omit small function words.
Wernicke's aphasia	Wernicke's aphasia is a general inability to understand speech, occurring following damage to the left temporal lobe. It consists of difficulty recognizing spoken words, difficulty understanding the meaning of spoken words, and difficulty converting thoughts into words. Speech is a mixture of real words, repeated words and neologisms, showing difficulties in retrieving words from memory.
Other aphasias	Conduction aphasia is an inability to repeat speech sounds unless they are meaningful, resulting from damage to the arcuate fasciculus. In transcortical sensory aphasia patients can repeat words and pseudowords spoken to them, but they cannot understand the meanings of words. This results from damage to the cortex surrounding Wernicke's area.
Acquired dyslexia	Surface dyslexia is an inability to recognize words from their physical appearance. Patients seem unable to access word meaning by the direct lexical access route. Patients with phonological dyslexia can read words by the direct lexical access route, but they are unable to translate written characters into sounds (the grapheme–phoneme route).
Developmental dyslexia	Developmental dyslexia is a range of difficulties with reading that develop in otherwise normal children. Most dyslexics have difficulties with both word-form and phonological reading. A variety of related deficits have been identified, although none applies to all dyslexics.
Writing and reading	We can produce a correctly spelled written word either by translating sounds (phonemes) into their written equivalents, or by imagining the written word and produce a matching word on the paper. In phonological dysgraphia a person cannot spell phonetically. Orthographic dysgraphia is a deficit in the ability to use visual processes in spelling.
Related topics	Hemispheric lateralization (C3) The neural basis of language (Q2) Brain damage and recovery (C4)

Language   Language is central to human social and cognitive processes. It is our main way
                      of representing knowledge and ideas that we manipulate in thinking. We use
                      language to communicate through speech, writing and signing. Much of our
                      understanding of the nature of language production and comprehension comes
                      from the study of individuals with specific disorders.

                      Aphasia in general is an impairment of language production or comprehen-
                      sion. Most cases of aphasia result from *cerebrovascular accidents* (strokes).
                      Agraphia is the inability to produce written language, and **dysgraphia** is an
                      impairment of written language. The two terms are often used interchangeably.
                      Dyslexia refers to disorders of reading. In the following Topics we will examine
                      the most important examples of each of these types of disorder.

Broca's aphasia   In the 19th century, several clinicians noticed that damage to the left side of the
                      brain frequently caused impairment of speech, while damage to the right side
                      rarely did. In 1861, Paul Broca reported on the post-mortem examination of the
                      brain of a patient who for many years had been unable to speak, although he
                      could understand speech and was able to communicate by signals. This condi-
                      tion is now called **Broca's** (or **expressive**) **aphasia**. After examining the brains
                      of several other patients with expressive aphasia, Broca noted that they all had
                      damage to an area of the left inferior frontal cortex. This area is now called
                      Broca's area (see *Fig. 1*). Damage to the equivalent area of the right cerebral
                      cortex generally has little or no effect on language. This **lateralization of func-
                      tion** is now viewed as fitting with the general observation that the left hemi-
                      sphere is specialized for activities involving *temporal* or *sequential* control, while
                      the right cortex is specialized for activities involving *spatial* functions (see Topic
                      C3), and perhaps *emotional* ones (Topic O3).

                      There are a wide range of deficits in patients with Broca's aphasia. Some
                      patients cannot speak at all, but others can speak with difficulty. They tend to
                      speak slowly, non-fluently, and with great effort, although they may be able to
                      sing fluently the lyrics of a familiar song. They often use the wrong word, or
                      mispronounce words, and are aware that they are making mistakes. They often

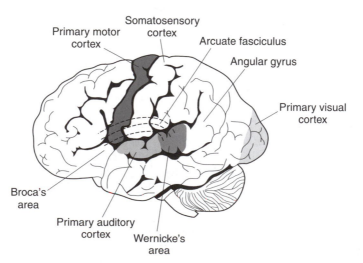

*Fig. 1.   Cortical regions involved in language production. The arcuate fasciculus is in the
white matter beneath the visible cortex.*

use only very simple grammatical constructions, and tend to omit small words with important grammatical functions, such as *a, the, in, by* or *some*. Language comprehension (see Topic Q2) is often unaffected for simple sentences, but more complex grammar, for example sentences in the passive voice, may cause difficulty. They may also have difficulty understanding written material, particularly with more complex grammar.

Wernicke's aphasia

Wernicke's aphasia (also known as **receptive** or **sensory aphasia**) is a general inability to understand speech, whether produced by other people or by oneself. The disorder is named after a German neurologist, Carl Wernicke, who described the condition in 1874. He showed that it occurs following damage to part of the left temporal lobe of the cerebral cortex, just behind the primary auditory cortex, now called **Wernicke's area** (see *Fig. 1*).

Wernicke's aphasia is not a simple disorder of speech recognition. It consists of at least three deficits:

- difficulty *recognizing* spoken words;
- difficulty *understanding the meaning* of spoken words;
- difficulty *converting* thoughts into words (*retrieval*).

People with Wernicke's aphasia produce a characteristic type of speech. This has apparently normal grammatical structure, function words (e.g. *the, of*), rhythm and intonation, but is a mixture of real words, repeated words and invented words (*neologisms*). For example, a patient asked to describe his job said:

> "I wanna tell you this happened when happened when he rent. His, his kell come down here and is, he got ren something. In these ropiers were with him for hi, is friend, like was … And he roden all o these arranjen from the peddis on from is pescid."
>
> (from Kertesz, 1981, p. 73)

The repeated and invented words are thought to show difficulties in retrieving words from memory. Patients speaking like this are usually unaware that they are not producing proper language. This contrasts with word retrieval difficulties that we may all experience, and which may occur more frequently with age.

Other aphasias

A number of other types of aphasia have been described. These include **conduction aphasia**, which is an inability to repeat speech sounds unless they are meaningful. Patients can speak and understand speech normally, and can repeat real words, but when asked to repeat pseudowords (pronounceable letter sequences that are not words in real language), are unable to do so. Sometimes, instead of repeating the spoken word, the patient will reply with a semantically related word (e.g. *car* for *auto*). Presumably, patients are responding not by repeating sounds, but by understanding the meaning of what is said, and producing a word with related meaning. Presented with alternative pronunciations, they can often choose the correct one. This suggests that the underlying deficit is not to do with the processing of speech sounds, but with retrieval (Goodglass, 1993). Wernicke suggested that conduction aphasia results from damage to a fiber bundle, the **arcuate fasciculus**, joining Broca's and Wernicke's areas (see *Fig. 1*). However, it has more recently been suggested that it can result from damage to the temporal cortex. Quigg and Fountain (1999), for example, produced the signs of conduction aphasia by stimulating the left posterior superior temporal gyrus in conscious epileptic patients.

Transcortical sensory aphasia is in a sense the converse of conduction aphasia. Patients can repeat words and pseudowords spoken to them, but they cannot understand the meanings of words, and they produce speech like that in Wernicke's aphasia. This condition results from damage to the cortex surrounding Wernicke's area. It seems to be a failure of transfer of information from non-language brain regions to the parts involved in encoding it into a semantic form (Kertesz, Sheppard and Mackenzie, 1982).

Acquired dyslexia

Acquired dyslexia refers to a number of specific forms of reading difficulty resulting from damage to specific regions of the brain. Taken together, these disorders suggest two routes by which word meaning is obtained from written words (Hinton, Plaut and Shallice, 1993). These routes are based on different brain mechanisms. Normally, reading involves both of them.

Surface dyslexia is an inability to recognize words from their physical appearance (*word-form reading*). That is, a patient can read and correctly pronounce words and pseudowords with regular spellings, but mispronounces (and may fail to recognize) irregular words, for example reading *yacht* as 'yatcht'. Patients thus seem unable to access word meaning by the *direct lexical access* route.

Conversely, patients with **phonological dyslexia** can read words by the direct lexical access route; that is they extract meaning directly from word form. But they are unable to translate written characters into sounds (the *grapheme–phoneme* route). Typically, they can read familiar real words (including irregularly spelt ones), but cannot sound words out, nor read unfamiliar words or pseudowords. The locations of lesions producing surface and phonological dyslexia are not clear, but the left temporal lobe has been implicated in both (Rosenberger, 1990).

Alexia is an inability to read. It may be observed in patients who have normal abilities in speaking and understanding speech following damage to the **angular gyrus**, a cortical area between the visual cortex and Wernicke's area.

Developmental dyslexia

The term **developmental dyslexia** (or commonly just *dyslexia*) is applied to a range of difficulties with reading that develop in children of normal intelligence and no obvious perceptual deficits. It tends to run in families, has important educational implications, and occurs more often in boys. About two-thirds of dyslexics have difficulties with both word-form and phonological reading. Of the remainder, about two-thirds have solely a phonological problem, and one-third a word-form problem. A variety of related (or underlying) deficits have been identified, although none of these seems to apply to all dyslexics. These include impaired verbal memory, difficulties with form perception (including discrimination and memory) and difficulties with maintaining visual fixation and with visually tracking moving targets.

Post-mortem examinations have shown abnormalities in the **magnocellular layers** of the **lateral geniculate nucleus** in the visual system (see Topic G1) in the brains of many dyslexics. Studies with fMRI scans (see Topic A2) have confirmed an abnormality in the magnocellular system. **Area V5** in the visual cortex, to which the magnocellular system projects, shows activity in normal people viewing moving targets, but not in dyslexics (Eden et al., 1996). The cerebral cortices of dyslexics show many small, widespread areas of displaced neurons, **ectopias**, which result from errors in neuronal development in the fetus (Galaburda et al., 1985). Using biochemical and imaging methods, other researchers have shown abnormalities in the **cerebellum** in developmental dyslexia (see Nicolson et al., 2001).

Writing and reading

Writing involves changing spoken language, which we learn first, into a conventional visual form. There are two ways in which we can produce a correctly spelled written word, which parallel the two ways of reading. One is to translate the sounds (phonemes) from which the word is constructed into their written equivalents. You can observe yourself doing this when writing a very long word, especially an unfamiliar one. The other is to imagine the written word and produce a matching word on the paper. In an irregular language like English, many people will write a word in more than one possible spelling and check which one looks right. Studies of people with specific difficulties with spelling show that different neural mechanisms underlie these two ways of writing.

Phonological dysgraphia (sometimes called **phonological agraphia**) is a disorder in which a person cannot spell phonetically (using word sounds). Patients may still, however, be able to use visual strategies to produce correct words. They cannot, however, write pseudowords. Phonological dysgraphia follows damage to the upper part of the temporal lobe (Benson and Geschwind, 1985).

In contrast, **orthographic** (or **surface**) **dysgraphia** is a specific deficit in the ability to use visual processes in spelling. Patients have particular difficulty with irregular words (e.g. *flight*, *yacht*). However, they have little trouble with more regularly spelled words, nor with pseudowords, both of which they spell phonetically. Orthographic dysgraphia is a consequence of lesions in the lower part of the parietal lobe (Benson and Geschwind, 1985).

Q2 THE NEURAL BASIS OF LANGUAGE

Key Notes

The Wernicke–Geschwind model	The Wernicke–Geschwind model has seven components: the primary auditory cortex, the primary visual cortex, the left angular gyrus, Wernicke's area, the arcuate fasciculus, Broca's area and the primary motor cortex. Damage to the components of the model produces specific disorders.
Problems with the Wernicke–Geschwind model	First, brain damage is rarely localized. The various symptoms and signs shown by different Broca's aphasia patients reflect different patterns of damage in and around Broca's area. Second, disorders of language are rarely simple and 'pure'. Third, highly localized surgical lesions do not have the expected effects on language. Fourth, electrical stimulation of the brains of conscious patients produces effects on language that are not expected on the basis of the model.
Language and the brain	Language involves many cortical areas and subcortical structures. Listening to words produces activity in the secondary auditory area. Repeating a word aloud adds activity in the primary motor cortex. Saying a word semantically associated with a heard word produces further activity in and around Broca's area. A similar pattern emerges with visually presented words. Several other areas are involved in language. Other imaging research has shown considerable variability in the areas activated by reading and saying sentences. These involve large areas of the frontal cortex, the upper parts of the temporal lobe, and part of the parietal lobe, all regions of the left cerebral hemisphere. The right hemisphere does have some language function, more so in less strongly right-handed people.
Related topics	Hemispheric lateralization (C3) Language and its disorders (Q1) Brain damage and recovery (C4)

The Wernicke–Geschwind model

Wernicke first suggested in the 19th century that Broca's area and Wernicke's area are two language centers, one concerned with speech *production* and the other with speech *perception*. He argued that disconnecting these two centers, by damage to the arcuate fasciculus, would produce just the pattern of difficulty seen in conductive aphasia (see Topic Q1).

Geschwind (1965) revived and developed these views, producing a model of the cortical basis of language: the **Wernicke–Geschwind model**. The model has been influential but, as we will see later, it is simplistic and inaccurate. It has seven major components:

- the **primary auditory cortex**, which receives speech as sensory projections;
- the **primary visual cortex**, which receives written language as sensory projections;

- the left **angular gyrus**, which receives information from the visual cortex;
- **Wernicke's area**, which receives inputs from the angular gyrus, and directly from the auditory cortex, and is responsible for language comprehension;
- the **arcuate fasciculus**, which transfers verbal information (e.g. thoughts) to **Broca's area**, which programs coordinated activity in the **primary motor cortex**, producing speech.

Damage to the components of the model is said to produce specific disorders as described in Topic Q1 (see *Table 1*).

Table 1. *Lesions and associated language disorders in the Wernicke–Geschwind model*

Location damaged	Disorder
Primary auditory cortex	Deafness
Primary visual cortex	Blindness
Angular gyrus	Alexia
Wernicke's area	Receptive (sensory) aphasia
Arcuate fasciculus	Conduction aphasia
Broca's area	Expressive aphasia
Primary motor cortex	Paralysis

The model proposes that language works as follows (see also *Fig. 1* of Topic Q1). Suppose you are answering questions spoken by another person. The question is heard by your *primary auditory cortex*, and transferred to *Wernicke's area*. Here, the question is comprehended. A response is generated (although the model has little to say about how), and is converted into linguistic form, also by Wernicke's area. This neural representation of the response is passed through the *arcuate fasciculus* to *Broca's area*, where it starts the coordinated process of speaking, the components of which are controlled by the *primary motor cortex*. The process for reading is similar, except that the written words are perceived by the *visual cortex*. Their neural representation is converted in the *angular gyrus* to the neural representation of the equivalent auditory input. This is then passed to *Wernicke's area*, as before.

Problems with the Wernicke–Geschwind model

There are numerous difficulties with the Wernicke–Geschwind model, and with the observations on which it is based. First, *the aphasias and other disorders of language described in Topic Q1 are rarely simple and 'pure'*. For example, cases of Broca's aphasia usually show aspects of receptive as well as expressive aphasia, and Wernicke's aphasia patients usually have expressive as well as receptive difficulties.

Second, *electrical stimulation of the brains of conscious patients produces effects on language that are not expected on the basis of the model*. Stimulation of brain sites widely distributed across the temporal, parietal and frontal lobes can disrupt speech production and comprehension. The effects on language are not restricted to stimulation of the areas named in the model (Penfield and Roberts, 1959).

Third, *brain damage is rarely restricted to clearly defined, local regions*. For example, Broca's own patients had brain damage extending well beyond Broca's area. The various symptoms and signs shown by different patients reflect different patterns of damage in and around Broca's area (see Penfield and Roberts, 1959). Patients with Broca's aphasia sometimes cannot obey a simple command such as to stick out the tongue. This suggested to some people that the underlying deficit is an inability to make the movements of speech. (Broca's area is immediately adjacent to the part of the primary motor cortex that controls movements of the lips and tongue.) However, it is apparent from the features of Broca's aphasia described in Topic Q1 that the deficit is in *language* rather than movement. Cases showing inability to move mouth parts following instructions must involve damage extending to the primary motor cortex.

Modern studies using imaging techniques have also shown that aphasic patients have much more extensive damage than suggested by the model, including damage to subcortical pathways (e.g. Naeser et al., 1981). What is more, lesions confined to Broca's area rarely produce *persistent* aphasia. Patients usually show some recovery after a few weeks of aphasia. Conversely, expressive aphasia can be produced by lesions that do not include Broca's area at all. A similar pattern of findings relates to receptive aphasia and Wernicke's area (Penfield and Roberts, 1959).

Language and the brain

The use of imaging methods has shown that the Wernicke–Geschwind model is too simple, but has not yet produced a clear alternative account of how language is organized in the brain. It is apparent, however, that language involves far more cortical areas, and subcortical structures, than the model proposed (see *Fig. 1*).

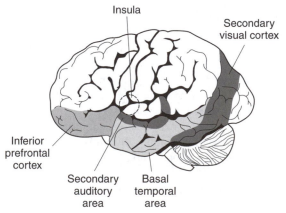

Fig. 1.   Other brain areas involved in language. The insula is out of sight between the temporal and frontal lobes.

Imaging studies have confirmed in intact, normal people the general picture obtained with lesion and stimulation studies on surgical patients. That is, language involves wider areas of the cortex than the traditional 'speech centers' (Binder, 1997). Briefly, silently listening to words produces activity in the upper temporal lobe, extending beyond Wernicke's area. This is the **secondary auditory area** (see Topic H1). Repeating a word aloud adds activity in the primary motor cortex. Saying a word semantically associated with a heard word (e.g. *eat* following *cake*) produces further activity in and around Broca's area. This suggests that Broca's area is involved with thinking, or converting thought into speech, rather than the act of speaking. A similar pattern emerges with visually presented words. Silently reading words produces activity in the **secondary visual cortex**, repeating them activates the primary motor cortex, and producing a word with an associated meaning further activates cells in and around Broca's area.

Studies of patients with relatively localized lesions in the **insula**, an area of cortex between the frontal and temporal lobes, show this to be involved in coordinating speech movements (Dronkers, 1996). Most cases of Broca's aphasia include damage to the insula. Other areas shown by imaging studies to be involved in aspects of language include the **left inferior prefrontal cortex**, which is activated during word memory retrieval tasks. Blocking activity in the **basal temporal area** disrupts a person's ability to name objects.

Other imaging research has shown considerable variability in the areas activated by reading and saying sentences. First, different brain sites are activated in different people by the same task. Second, the area activated by a task varies from trial to trial in the same person. Added together, these involve large areas of the frontal cortex, the upper parts of the temporal lobe, and part of the parietal lobe.

All of these findings, and the lesion and stimulation observations mentioned previously, predominately relate to regions of the *left* cerebral hemisphere. The right hemisphere is not completely devoid of language function, however, even in the most strongly right-handed people. 'Split-brain' patients (see Topic C3), who have the left and right hemispheres disconnected, cannot recognize words presented to the left visual half field (hence to the right hemisphere alone). However, if a word is presented briefly to the right hemisphere, the left hand (controlled by the right hemisphere) can pick out the correct object from among several hidden behind a screen. Thus, the right hemisphere is able to understand the word and direct the left hand to locate the appropriate object, without any information reaching the left hemisphere. The person, of course, would not be able to say what the object is since speech production is clearly controlled in the left hemisphere (Gazzaniga, Bogen and Sperry, 1962).

While the left hemisphere dominance for language is very clear in strongly right-handed people, with increasing degrees of left-handedness (**sinistrality**) more language abilities are located in the right hemisphere. For example, while about 60% of right-handed patients with left hemisphere lesions have aphasia, only about 30% of left-handers do. Conversely, only 2% of right-handers with right hemisphere damage are aphasic, while about 25% of left-handers with similar damage are. Temporarily stopping the activity of one hemisphere by the **Wada test** (see Topic C3) confirms this. Milner (1974) showed that over 90% of right-handers had speech located in the left hemisphere, while about 70% of left-handers were left-hemisphere dominant for speech.

R1 SCHIZOPHRENIA

Key Notes

The nature of schizophrenia	Schizophrenia is characterized by loss of contact with reality. Schizophrenics may have hallucinations, delusions, may shut themselves off from the world, have poverty of speech, blocking of thought, formal thought disorders, psychomotor symptoms including catatonia, inappropriate affect, or blunted affect. Type I schizophrenia shows predominantly the 'positive symptoms' of delusions, hallucinations and disorganized speech. Type II schizophrenia has the 'negative symptoms' of flat affect, poverty of speech and reduced voluntary movement. Schizophrenia is related to socioeconomic and marital status. Sociocultural stress, physical stress and unusual communication patterns in families have been proposed as factors.
The genetics of schizophrenia	It is likely that several genes are involved in predisposing people to schizophrenia. These genes control aspects of neural or endocrine development. Research has identified several locations on a number of chromosomes that appear to be linked with schizophrenia, but they are often not confirmed in other samples. Candidate genes are involved in aspects of synaptic function, or with cannabinoid receptors.
The psychopharmacology of schizophrenia	According to the dopamine hypothesis, the underlying pathology in Type I schizophrenia is an increase in activity in central dopaminergic circuits. The early antipsychotic drugs acted by blocking dopamine receptors. The neurons affected extend from the midbrain into the nucleus accumbens and the amygdala, areas that are involved in the effects of reinforcement and with linking perception with emotion and memory. Excess dopamine activity may cause the brain to be unable to ignore sensory inputs or internally generated thoughts, leading to a sort of 'cognitive overload'.
The brain and schizophrenia	The negative symptoms seen in schizophrenia are similar to those often seen following brain damage. The brains of people with Type II schizophrenia frequently have enlarged cerebral ventricles, particularly the third ventricle. Abnormalities have been found in the medial temporal lobes (including the amygdala and hippocampus) and the frontal lobes. These abnormalities might be directly attributable to genetic factors, or to damage from environmental factors.
Related topics	Psychoactive drugs (F3) Mood disorders (R2)

The nature of schizophrenia

Schizophrenia (literally 'split mind') is a term introduced by Eugen Bleuler in 1911 to describe a particular psychotic condition (or group of conditions). Schizophrenia does *not* involve multiple personality; the term implies a split with reality. About 1% of people meet the criteria for schizophrenia; men and women about equally (American Psychiatric Association, 1994). The incidence is nearly five times as high in the lowest socioeconomic groups as it is in the highest and is about three times higher in separated or divorced people, and twice as high in single people, as in married people. In the UK and the USA the incidence is much higher in people of Afro-Caribbean origin than in whites (Keith, Regier and Rae, 1991). It is difficult to establish the direction of cause and effect in the relationships between incidence and socioeconomic and marital status. The stress of poverty or marital breakdown may contribute to the onset of schizophrenia, but on the other hand schizophrenic behavior may be a factor causing poverty (e.g. through difficulty keeping jobs) and marital discord.

Schizophrenia is characterized by loss of contact with reality. This may involve such extreme distortion of perception and thought that the person is unable to function adaptively. The symptoms are highly variable. Schizophrenics may have **hallucinations** (false sensory perceptions – usually hearing voices), or **delusions** (false beliefs), or may shut themselves off from the social and physical world. A person may have one overwhelming delusion (e.g. that he is Christ), or many delusions, and may feel uplifted by or confused by the delusions. Some people with schizophrenia show *poverty of speech*, characterized by brief and content-free replies. This may be accompanied by *blocking of thought* (having fewer and less varied thoughts than most people). On the other hand, some patients show positive *formal thought disorders*, including disorganized speech, following unexpected chains of thought, using invented words, and perseveration (the repetition of words or phrases).

Other patients show *psychomotor symptoms*, including loss of spontaneous movement, or the adoption of odd postures or movements. In extreme form, this may be a total lack of response to the environment: **catatonia**. Others show *inappropriate affect* (for example laughter when hearing tragic news) or *blunted affect*.

Increasingly, older classifications of subtypes of schizophrenia are being replaced by a simpler classification (see Andreassen, 1995). **Type I schizophrenia** shows predominantly the 'positive symptoms' of delusions, hallucinations and disorganized speech. (These are called positive as they are the *presence* of unusual characteristics.) **Type II schizophrenia** has the 'negative symptoms' of flat affect, poverty of speech, and reduced voluntary movement (called negative because of the absence of normal characteristics). This classification seems to fit better with certain brain dysfunctions (see below).

Schizophrenia is generally regarded as multifactorial in origin. The onset and course of the illness is influenced by sociocultural factors. One interpretation of the relationships between incidence and various socioeconomic factors is that these factors cause stress, which triggers schizophrenic symptoms as an escape from sociocultural pressures. Particular attention has been paid to the role of families. Bateson et al. (1956) proposed the **double-bind hypothesis**. The parents of schizophrenic patients were said to behave towards them in self-contradictory ways (double-binds), for example making positive statements with negative nonverbal signals. Repeated exposure to these ambiguous messages could lead to the development of abnormal ways of coping with the social environment. More recently, attention has turned to emotional communication within the family. Brown et al. (1972) showed that schizophrenic patients returning home to live with relatives are more likely to relapse if they have a relative who is high in **expressed emotion** (EE). This reveals itself in interviews as high levels of critical comment, hostility and/or emotional overinvolvement.

However, this is not a cause of schizophrenia, and it is not specific: EE has been shown to influence the course of several other conditions.

The physical environment influences the onset of schizophrenia. More schizophrenics are born in the first third of the year in the northern hemisphere (see *Fig. 1a*), but the last third in the southern hemisphere (Kendell and Adams, 1991). This has suggested that viral infections, most common in the winter, are somehow involved in predisposing a person to illness, perhaps by damaging the developing brain of the fetus. This is supported by the observation that the incidence of schizophrenia is particularly high in people born a few months after a flu epidemic (Sham et al., 1992; see *Fig. 1b*). The importance of prenatal

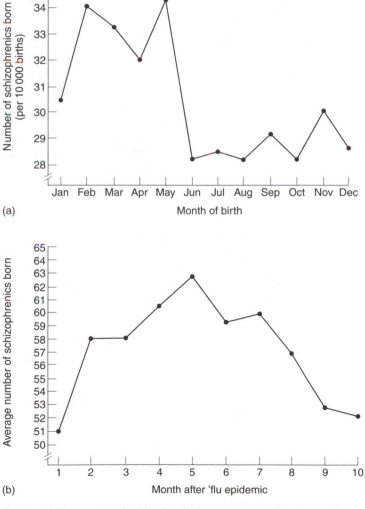

(a)

(b)

Fig. 1. (a) The number of schizophrenics born is greater after pregnancies during the winter months than after those during spring and summer. (Data from R.E. Kendell and W. Adams, British Journal of Psychiatry, 1991, 158, 758–763.) (b) More schizophrenics are born when a 'flu epidemic occurred during pregnancy. (Data from P.C. Sham et al., British Journal of Psychiatry, 1992, 160, 461–466.)

factors is supported by evidence that infants who later become schizophrenic show certain behavioral abnormalities soon after birth, particularly in movement and facial expression.

The genetics of schizophrenia

That there is a **genetic** component in schizophrenia is shown by several types of data.

The closer one's relationship with a schizophrenic person, the more likely one is also to be schizophrenic (see Gottesman, 1991). For first cousins, this is 2%; for fraternal twins 10–20%; for identical twins 45–50%. Nearly 50% of the children of two schizophrenic parents become schizophrenic. These figures do not prove a genetic causation, as similarities in child-rearing and other environmental factors are likely to vary in the same way with closeness of relationships (e.g. identical twins are often treated identically). However, the families of schizophrenics who had been adopted in infancy have been examined. Blood relatives of such schizophrenics are 4–5 times as likely to be schizophrenic as are members of the families they were adopted into. Taken together, these findings show that genetic factors are important.

It is likely that several genes are involved, and that the genetic factors predispose a person to develop schizophrenia under certain stressful conditions. These genes would control aspects of neural or endocrine development. Research has identified several locations on a number of chromosomes that appear to be linked with schizophrenia, but they are generally not confirmed in other samples. This is partly because a number of genes are involved and these, like the symptoms, will vary from patient to patient. Most recently, the following possible gene linkages have been reported:

- a gene involved in the formation of synapses, **dysbindin**, has been shown to be more variable in schizophrenics (Straub et al., 2002), although another group failed to confirm this (Morris et al., 2003). This again could lead to disruptions within the brains of schizophrenics;
- another gene, **COMT**, has been found by several research groups to be less active in some schizophrenics showing negative cognitive symptoms. The biochemical effect of the COMT gene is to produce an enzyme that breaks down dopamine in synapses. The occurrence of its weaker form coincides with decreased activity in the prefrontal cortex, and this in turn is associated with schizophrenic symptoms (see Weinberger et al. 2001);
- a gene that codes for a cannabinoid receptor, **CNR1**, is linked to schizophrenia with negative emotional and personality symptoms. This receptor responds to the active ingredient of cannabis (Ujike et al, 2002). Cannabis use in teenagers increases the incidence of later schizophrenia by a factor of six, and in the short term it can induce psychotic states.

The psycho-pharmacology of schizophrenia

At least some of the symptoms of schizophrenia may be produced by a neuroendocrine disturbance. According to the **dopamine hypothesis**, the underlying pathology is an increase in activity in dopaminergic circuits in the brain (Crow, 1982; see Topic B3). This was suggested first by the observation that antipsychotic drugs such as **chlorpromazine** often produce tremors like those in Parkinson's disease. Parkinson's disease results from insufficient dopamine in motor neural circuits. Drugs that increase dopamine activity, such as amphetamines, increase schizophrenic symptoms, and can produce psychotic symptoms in non-schizophrenics. The early antipsychotic drugs acted by blocking dopamine receptors. The symptoms on which these drugs work, and the effects of manipulating dopamine in the brain, are mostly positive symptoms. So, the hypothesis is that Type I schizophrenia results from overactivity of dopaminergic neurons.

The neurons affected extend from the midbrain into the **nucleus accumbens** and the **amygdala**, areas that are involved in the effects of reinforcement (see Topic P2) and with linking perception with emotion and memory (see Topic O2). There is conflicting evidence about whether synapses in these areas produce more dopamine or are more active in schizophrenics. However, there is clearer evidence from *post mortem* studies that the brains of schizophrenics contain more of two particular types of dopamine receptor: D_3 and D_4. Gurevich et al. (1997), for example, showed that there were twice as many D_3 receptors in the basal ganglia and ventral forebrains of untreated schizophrenics as in treated schizophrenics and non-schizophrenics. However, the most effective antipsychotic drugs act mostly on a third receptor type, D_2, and this is generally supposed to be the receptor whose function is increased in schizophrenia. *Low* levels of dopamine have been reported in the prefrontal cortex of people with negative schizophrenic symptoms (Davis et al., 1991). This could explain why positive and negative symptoms can occur together. Exactly how these disturbances of dopaminergic synapses produce symptoms is unclear. It may be that excess dopamine activity causes a state in which the brain is unable to ignore sensory inputs or internally generated thoughts, leading to a sort of 'cognitive overload'.

Evidence that other endocrine systems might be involved in producing schizophrenic symptoms comes from the anti-psychotic actions of drugs which act on other receptors, such as serotonin and glutamate (Butini et al., 2003). This emphasizes the current uncertainty about the dopamine hypothesis and the complex nature of schizophrenia, and also suggests that we should look at neural structures as well as endocrine systems.

The brain and schizophrenia

Although pioneers in the field of schizophrenia in the early 20th century, such as Bleuler and Kraepelin, believed schizophrenia results from an organic brain disorder, this underlying pathology is still unclear. The negative symptoms seen in schizophrenia are similar to those often seen following brain damage, and are often accompanied by clear neurological symptoms and signs, such as altered reflexes and eye movement peculiarities. *Post mortem* examinations of the brains of people with Type II schizophrenia revealed that they frequently had enlarged cerebral ventricles (see Topic C1), particularly in the right hemisphere. A review of imaging studies (Shenton et al., 2001) reported that 80% of studies find ventricular enlargement, mostly of the **third ventricle**. The third ventricle is in the diencephalon, adjacent to the thalamus and hypothalamus. Abnormalities of the **medial temporal lobes**, including the **amygdala**, the **hippocampus** and **temporal cortex**, were found in 74% of studies, and 67% reported frontal lobe involvement, particularly **prefrontal** and **orbitofrontal** regions. These regions are associated with various emotion, memory and language functions, and this could account for the variety of symptoms in schizophrenia.

While some of these abnormalities might be directly attributable to genetic factors, another possibility is that they result from a genetically-based susceptibility to damage from environmental factors. These environmental factors would include viruses, such as might be implicated by the incidence of schizophrenia in relation to date of birth reviewed earlier.

R2 MOOD DISORDERS

Key Notes

Major depressive disorder	Major depressive disorder is characterized by sadness, fatigue, physical symptoms, sleep and psychomotor disturbances, reduced drive, poor appetite and concentration, feelings of worthlessness and self-blame, and thoughts of death. Depression is often triggered by stress. Both cognitive and neurochemical factors are involved, and there are genetic factors.
The psychopharmacology of depression	The monoamine hypothesis proposes that depressive illness results from a deficit in monoamine neurons. Some drugs lower monoamine levels and cause depression, others increase monoamines and relieve depression. However, reduced amounts of the breakdown products of monoamines are not always found in patients' urine; no decrease in monoamine release has been found in the brains of depressed people; the biochemical effect of tricyclic antidepressants is immediate, but their clinical effect on symptoms is delayed. Alternatives to the monoamine hypothesis are reduced activity in the central reward circuits and deterioration of the response of the hypothalamic–pituitary–adrenal axis.
The brain and depression	The most frequently reported abnormalities are in the limbic system, basal ganglia, frontal lobes, hippocampus, amygdala and prefrontal cortex.
Bipolar disorder	In bipolar disorder the individual swings between extremes of serious depression and mania. There is a strong genetic component. Attempts to extend the monoamine hypothesis to bipolar illness have not been successful. However, serotonergic synapses are likely to be involved. The mode of action of the main treatment for mania, lithium, is unclear.
Sleep and depression	Depressed people tend to have less slow-wave sleep and more rapid eye movement (REM) sleep. Sleep deprivation reduces the severity of depression. Most classes of drug that relieve depression reduce REM sleep and increase slow-wave sleep. Seasonal affective disorder occurs in the winter and is characterized by depression, fatigue, *increased* sleep and *increased* appetite. It may be treated by phototherapy.
Related topics	Disruptions of sleep and rhythms (K5) Anxiety disorders (R3) Central mechanisms in emotion (O2)

Major depressive disorder

The affective disorders we consider here go far beyond the fluctuations of mood we all experience from time to time. There is considerable debate about how many different disorders exist. Those we will consider are debilitating clinical states, often with psychotic features. The most frequent is **major depressive disorder** (sometimes called **unipolar disorder**). In any one year, about 5% of the population of developed countries suffers from major depressive disorders.

Twice as many women as men are diagnosed with the condition. It is more frequent in separated and divorced persons than in those of other marital status (Weissman et al., 1991).

The characteristic symptom of major depression is persistent, often extreme, sadness (depressed mood). Other symptoms include fatigue, varied and often vague physical symptoms, sleep disturbances, psychomotor disturbances (decreased, or sometimes increased, activity levels), reduced drive, poor appetite, reduced ability to concentrate, feelings of worthlessness and self-blame, and frequent thoughts of death. Sufferers are at risk of suicide, but less often when in the deepest despair than when showing signs of recovery. Women attempt suicide about three times as often as men, but men are four times as likely to succeed (see McIntosh, 1991).

Episodes of depression often seem to be triggered by stressful events, and those subject to high levels of stress are more likely to develop depression. Both cognitive and neurochemical factors are involved in depressive illness. There is evidence from studies of twins and other relatives, and from adoption studies, that there are genetic factors in unipolar depression (Sullivan, Neale and Kendler, 2000). For example, about 50% of identical twins of a sufferer will also suffer. It is expected that multiple genes are involved. At present, however, little progress has been made towards locating the genes, although relevant genes are likely to be those affecting monoamine synapses and other neural structures.

The psycho-pharmacology of depression

The **monoamine hypothesis** of affective disorders proposes that depressive illness results from a deficit in neurons using the monoamines **norepinephrine** and **serotonin** as transmitters. Numerous lines of evidence support this theory (see Leonard, 2000):

- **reserpine**, a drug used to treat hypertension, eventually *lowers* norepinephrine levels and sometimes causes depression. Its action is to prevent the take up of monoamines into synaptic vesicles, reducing the amount released into synapses (see Topic F2);
- antidepressant drugs called **monoamine oxidase inhibitors** (**MAOI**) work by increasing the availability of norepinephrine in synapses, by inhibiting the action of the enzymes that break it down;
- another class of antidepressants, **tricyclics**, increase the amount of both norepinephrine and serotonin in synapses by blocking their reuptake;
- a newer class of antidepressants, the **selective serotonin reuptake inhibitors** (**SSRIs**), including Prozac, increase the amount of serotonin in serotonergic synapses, but do not affect noradrenergic synapses;
- the cerebrospinal fluid (see Topic C1) of suicidal patients contains less of a breakdown product of serotonin (**5-HIAA**). This shows that less serotonin is being released in the brains of these patients;
- *post mortem* studies show increased numbers of some subtypes of serotonin and norepinephrine receptor in the brains of people who had had depressive illness (Nemeroff, 1998). This is interpreted as a compensatory change following lower release of these neurotransmitters in these patients.

There are some difficulties with the monoamine hypothesis of depression. We would expect to see decreased amounts of the breakdown products of monoamines in the urine of patients, but this is not consistently the case. Furthermore, although as we have just seen the increase in the number of monoamine receptors suggests decreased monoamine release, no such decrease has been found in the *post mortem* brains of depressed people. The third problem is that the biochemical effect of tricyclic antidepressants is immediate, but their clinical effect on depressive symptoms develops only after 2–3 weeks. A possible explanation of this is that pre-synaptic neurons have **autoreceptors**; receptors for their own transmitter on their own pre-synaptic membrane. The function of these is to control transmitter release. The slow clinical action of antidepressants might result from the gradual adaptation of these autoreceptors (see Nutt, 2002). Either they become less sensitive, or they decrease in numbers in response to the increased amount of monoamines in the synaptic cleft resulting from the inhibition of reuptake by the antidepressant.

Alternatives to the monoamine hypothesis have been proposed. Naranjo, Tremblay and Busto (2001) have reviewed evidence suggesting that reduced activity in the central reward circuits (see Topic P2) is involved in producing major depressive illness. Others have proposed that deterioration of the response of the hypothalamic–pituitary–adrenal axis, which we looked at in stress (Topic O5), is associated with major depression (see Hatzinger et al., 2002).

The brain and depression

Various abnormalities in the brains of people with affective disorders have, in recent years, been reported using a variety of methods. A review of PET imaging studies (Videbech, 2000) has shown that the most frequently reported regions showing abnormalities are the **limbic system**, the **basal ganglia** and the frontal lobes. Other studies have revealed reduced size of the **hippocampus** (Frodl et al., 2002; see Topic P2), and increased activity and blood flow in the **amygdala** and in the **prefrontal cortex** (Drevets, 1999).

Bipolar disorder

In the **bipolar disorders** the individual swings between extremes of serious depression and **mania** (the condition was formerly called manic–depressive illness). Manic episodes are periods of persistent elevated or irritable mood, with inflated self-esteem, decreased sleep, being extremely talkative, distractibility, 'flights of fancy', agitation, or engaging in dangerous activities. The rate at which a person swings from one extreme to another can vary between a few hours and several months, and the manic phase can last for days or months, but is typically 2–3 weeks long. A person might have only one manic or depressed episode. Bipolar disorders are less frequent than unipolar disorders: about 0.5–1% in the population. Men and women are equally likely to be affected (Weissman et al., 1991).

The causation of bipolar illness has a greater genetic component than unipolar illness (Kelsoe, 2003). 70% of the identical twins of sufferers also have bipolar illness. Again, little progress has been made towards locating the genes involved. Kelsoe suggests that most likely different populations of genes predispose people to overlapping varieties of bipolar illness.

Attempts have been made to extend the monoamine hypothesis to bipolar illness, but without much success. Norepinephrine is low in episodes of depression, and higher than normal in manic episodes (Post, Ballenger and Goodwin, 1980). Reserpine (see above) sometimes decreases manic symptoms. The most effective treatment for bipolar illness, **lithium** salts, has a rapid effect when administered in the manic phase, and also prevents the next swing to depression. Lithium has many biochemical effects, but its mode of action in bipolar illness is still unknown (Shaldubina et al., 2001). It inhibits serotonin autoreceptors, increasing the effectiveness of serotonergic synapses. Serotonin activity is *reduced* in both mania and depression. The drug **tryptophan**, which increases serotonin, can alleviate the symptoms of manic patients, suggesting that a low level of serotonin is a causal factor. In the longer term, lithium increases reuptake of glutamate. This reduction in excitatory activity could be responsible for at least part of its anti-manic effect.

Sleep and depression

Sleep disturbance is a very common symptom of affective disorders (see Topic K5). A depressed person tends to have less slow-wave (SW) sleep, more Stage 1 sleep, more rapid eye movement (REM) sleep, and to awaken more frequently. Sleep deprivation, particularly of REM sleep, reduces the severity of depression (Vogel et al., 1980). However not all patients benefit, and the depression returns after the first full night's sleep (Wu and Bunney, 1990; see *Fig. 1*). Most classes of drug that relieve depression reduce the amount of REM sleep, and increase the amount of SW sleep. Why this link between sleep and depression occurs is unknown (Van Bemmel, 1997).

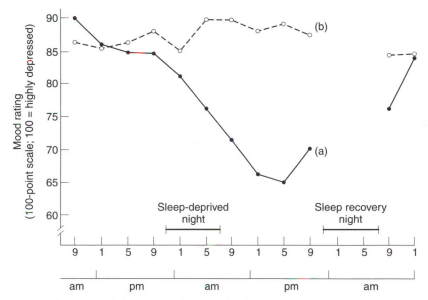

Fig. 1.   The effects of sleep deprivation on self-ratings of depression. Patients whose mood improves during the day (a) show further improvement when sleep-deprived. After the next night's sleep the benefit is lost. Those whose depressed mood continues through the day (b) show no such benefit. (Data from Wu and Bunney, 1990.)

Another link between depression and sleep is seen in **seasonal affective disorder (SAD)**. This illness is distinct from major affective disorder. It occurs in the winter months in a small number of people, and is characterized by depression, fatigue, *increased* sleep, and *increased* appetite and consequent weight gain. It is believed that this condition is associated with a disturbance of the process by which melatonin acts on the suprachiasmatic nucleus to synchronize diurnal rhythms (see Topic K4). It may be treated by **phototherapy**: exposure to bright light for several hours each day reduces symptoms in SAD sufferers, but not in other depressed patients (Terman et al., 1996). SAD might be caused by a failure of the body to synchronize with the day–night cycle because of the lower light levels of the winter. However, it does not matter at what time of day the phototherapy is administered (if it were resynchronizing the body clock it would be expected to be most effective in the early morning).

R3 ANXIETY DISORDERS

Key Notes

Panic attacks	Panic attacks are short periods of intense fear. Panic disorder consists of repeated panic attacks. There is a strong genetic influence. The condition depends on activity in noradrenergic pathways between the locus coeruleus and the amygdala. Various drugs that produce panic attacks in humans produce activity in the locus coeruleus. Drugs that increase norepinephrine activity trigger panic attacks in people with panic disorder, while drugs that decrease norepinephrine reduce panic symptoms. These findings suggest that panic attacks have at their core overactivity of noradrenergic neural circuits in the amygdala. Benzodiazepines, which reduce panic attacks, increase the sensitivity of the $GABA_A$.
Obsessive–compulsive disorder	Obsessions are persistent thoughts or images that intrude themselves into consciousness. Compulsions are repetitive behaviors that a person feels compelled to perform. Obsessive–compulsive disorder may start with strategies that dispel anxiety generated by unwanted thoughts. Imaging of sufferers' brains have shown increased activity in the orbitofrontal cortex and associated subcortical areas.
Other anxiety disorders	Generalized anxiety disorder is unfocused, prolonged anxiety and worry. There may be insufficient GABA activity in neurons connecting midbrain centers to the amygdala to inhibit fear responses. Posttraumatic stress disorder is persistent anxiety, intrusive thoughts, flashbacks, recurrent dreams and feelings of re-enactment of an intensely traumatic event. It may be more likely in people exposed to childhood traumas, or who lack a strong social support network. Phobic disorders are intense, irrational fears of specific objects or situations. The preparedness theory holds that phobias are most likely certain classes of (dangerous) natural stimuli, for which we have evolved a predisposition.
Related topic	Central mechanisms in emotion (O2)

Panic attacks Fear or anxiety experienced in physically or socially dangerous situations are normal, adaptive responses. When they are excessively prolonged or are unusually intense in relation to events, or occur with no precipitating event, then they might be considered pathological. A number of anxiety disorders are recognized, and a feature of many of them is **panic attacks**. A panic attack is a period of intense fear or discomfort commencing suddenly, and peaking within 10 minutes. It will have many of the following symptoms: palpitations or fast heart rate, sweating, trembling, sensations of difficulty breathing, chest discomfort or pain, nausea, dizziness or light-headedness, feelings of unreality, fear of losing control or dying, numbness or tingling, chills or hot flushes. If panic attacks occur repeatedly, with the person worrying between attacks that another will occur, a diagnosis of **panic disorder** might be made. This can be with or without **agoraphobia** (fear and avoidance of places from which escape might be difficult, or in which the person fears they might suffer a panic attack).

Family studies show that there is a powerful genetic influence in the origin of panic disorder (Weissman, 1993). Some 30% of the siblings of a sufferer will also be sufferers, and the incidence is much greater in identical twins. MacKinnon et al. (2002) reported that 90% of first degree relatives of people with panic disorder had a mood disorder, usually bipolar disorder, suggesting a shared genetic origin. Panic attacks can be triggered in susceptible people by any action that increases the activity of the sympathetic nervous system. This includes taking stimulants (e.g. coffee), exercise, and holding the breath (or otherwise increasing the carbon dioxide level of the blood).

The condition can be treated with some antidepressant drugs known to reduce norepinephrine activity (Rickels et al, 1993; see Topic R2). Major noradrenergic pathways link the **locus coeruleus** with the **amygdala**. The amygdala is known to be integral to the learning and triggering of emotions, particularly fear (see Topic O3). Various drugs that produce panic attacks in humans produce activity in the locus coeruleus (Singewald and Sharp, 2000). Drugs which specifically increase norepinephrine activity trigger panic attacks in people with panic disorder, while drugs that decrease norepinephrine reduce panic symptoms (Versiani et al., 2002). Taken together, these findings suggest that panic attacks have at their core overactivity of noradrenergic neural circuits that stimulate fear response mechanisms in the amygdala.

The other main class of drugs used to treat panic attacks, often in combination with behavioral techniques, is the **benzodiazepines** (e.g. Valium, Librium). These attach to part of the $GABA_A$ receptor, increasing its sensitivity to the neurotransmitter **GABA** (see Topic B3). This increases the inhibitory action of GABA. $GABA_A$ receptors are found widely in the brain, especially in regions known to be concerned with emotion, including the amygdala and the basal ganglia (see Topic O2). This suggests that some deficit in the GABA receptor system is at least partly responsible for panic attacks, presumably failing to inhibit adequately activity in fear circuits.

Obsessive–compulsive disorder

Obsessions are persistent thoughts or images that intrude themselves into consciousness. They are repetitive, uncontrollable, feel 'alien' to the person experiencing them (although they are recognized as coming from within), and cause anxiety. Attempts to dispel them or ignore them fail, and make the sufferer more anxious. **Compulsions** are repetitive behaviors or mental acts that a person feels compelled to perform. Failing to perform them, or trying not to perform them, leads to anxiety, so they seem to function to reduce anxiety. Common compulsions include hand-washing, checking (e.g. that doors are locked or electric appliances switched off), and counting. When extreme, obsessions or compulsions (usually both together) are diagnosed as **obsessive–compulsive disorder**. The condition can be extremely disabling, as both obsessions and compulsions can make it impossible to pursue a normal life.

PET scans of sufferers' brains have shown increased activity in the orbitofrontal cortex and associated subcortical areas. This activity decreases with successful treatment (Saxena et al., 1999). When sufferers were induced to hold apparently 'contaminated' items, fMRI showed that their agitation was accompanied by increased activity in various brain regions, including parts of the cortex (orbitofrontal, lateral frontal, insular and anterior cingulate) and the amygdala (Breiter et al., 1996). Conversely, recovery after treatment is accompanied by a decrease in prefrontal activity as measured by fMRI. All of these areas are involved in fear circuits.

Other anxiety disorders

When anxiety is prolonged (at least 6 months), but is not focused on particular events ('free-floating anxiety'), and is accompanied by difficulty concentrating, fatigue, feelings of inadequacy, insomnia or irritability, it may be diagnosed as **generalized anxiety disorder**. Up to 6% of the population may suffer from this in any year, and it is not as disabling as other anxiety disorders. Most sufferers, however, will at some time experience one of the acute anxiety disorders, or sometimes depression. Little is known of the causality of this disorder, and none of the proposed explanations is satisfactory.

There is little evidence of a genetic component in generalized anxiety disorder. The **benzodiazepines** (e.g. Valium, Librium), which we saw above work by attaching to GABA$_A$ receptors in the limbic system, are effective in treatment. We can postulate that the underlying circuits are the same as for panic disorder.

Shortly after a traumatic event, such as combat, rape, natural disaster or serious accident, during which a person has experienced intense fear, most people will experience anxiety, and may have recurrent intrusive thoughts about the event. In some people these persist long after, and are now considered to constitute **posttraumatic stress disorder** (**PTSD**). Sufferers experience vivid reliving of the events ('flashbacks'), may have recurrent dreams about the event, may feel or act as if the event were recurring, and are distressed by stimuli that remind them of the event. During each of these types of occurrence, they may experience heightened physiological arousal and its effects on behavior (e.g. irritability, lack of concentration, insomnia). Hippocampal lesions have been seen in some war veterans suffering from PTSD (Gurvits et al., 1996). A number of studies demonstrate a genetic component to PTSD. For example, Stein et al. (2002) have shown that monozygotic twins of people who suffer PTSD are more likely to suffer similar symptoms than are dizygotic twins. Segman et al. (2002) found that a gene coding for a molecule concerned with dopamine reuptake is mutated more frequently in PTSD sufferers that in those who do not react in this way to stress.

Phobias are intense and irrational fears of specific objects or situations. The object could be almost anything, commonly high places (acrophobia), spiders (arachnaphobia), and enclosed spaces (claustrophobia). While most of us might be apprehensive faced with many of these classes of object, in phobias the fear is pervasive and debilitating, including the symptoms of panic attacks. It may prevent a person leaving the house for fear of encountering the phobic object. Agoraphobia, which we mentioned in the preceding section, is more complex, with more general fears.

It is widely believed that specific phobias result from learned associations between the physical sensations of fear and particular objects which become the phobic object. Later, the fear sensations can **generalize** to other stimuli. One problem with this view is that phobias for a lot of dangerous objects (e.g. knives, cookers) are extremely rare. The **preparedness theory** (Seligman, 1971) suggests that we have evolved a readiness to respond with fear to certain classes of (dangerous) natural stimuli, making it more likely that phobias will develop to those.

REFERENCES

Adolphs, R. and Tranel, D. (1999) Intact recognition of emotional prosody following amygdala damage. *Neuropsychologia*, **37**, 1285–1292.

Adolphs, R., Damasio, H., Tranel, D., Cooper, G. and Damasio, A.R. (2000) A role for somatosensory cortices in the visual recognition of emotion as revealed by 3D lesion mapping. *Journal of Neuroscience*, **20**, 2683–2690.

Adolphs, R., Baron-Cohen, S. and Tranel, D. (2002) Impaired recognition of social emotions following amygdala damage. *Journal of Cognitive Neuroscience*, **14**, 1264–1274.

Albert, D.J., Dyson, E.M., Petrovic, D.M. and Walsh, M.L. (1988) Cohabitation with a female activates testosterone-dependent social aggression in male rats independently of changes in serum testosterone concentration. *Physiology and Behavior*, **44**, 9–13.

Albert, D.J., Petrovic, D.M. and Walsh, M.L. (1989) Competitive experience activates testosterone-dependent social aggression toward unfamiliar males. *Physiology and Behavior*, **45**, 723–727.

Aldrich, M.S. (1992) Narcolepsy. *Neurology*, **42**, 34–43.

Alexander, F. (1950) *Psychosomatic Medicine: Its Principles and Applications*. New York: Norton.

Alexander, G.M. and Sherwin, B.B. (1993) Sex steroids, sexual behavior, and selection attention for erotic stimuli in women using oral contraceptives. *Psychoneuroendocrinology*, **18**, 91–102.

Allen, L.S. and Gorski, R.A. (1992) Sexual orientation and the size of the anterior commissure in the human brain. *Proceedings of the National Academy of Science*, **89**, 7199–7202.

American Psychiatric Association (1994) *Diagnostic and Statistical Manual of Mental Disorders*, 4th edn. Washington, DC: APA.

Anand, B.K. and Brobeck, J.R. (1951) Localization of a 'feeding center' in the hypothalamus of the rat. *Proceedings of the Society for Experimental Biology and Medicine*, **77**, 273–275.

Andreassen, N.C. (1995) Symptoms, signs, and diagnosis of schizophrenia. *Lancet*, **346**, 477–481.

Archer, J. (1994) Testosterone and aggression. *Journal of Offender Rehabilitation*, **5**, 3–25.

Aschoff, J. (1994) Naps as integral parts of the wake time within the sleep/wake cycle. *Journal of Biological Rhythms*, **9**, 145–155.

Aserinsky, E. and Kleitman, N. (1953) Regularly appearing periods of eye motility and concomitant phenomena during sleep. *Science*, **118**, 273–274.

Aston-Jones, G. and Bloom, F.E. (1981) Activity of norepinephrine-containing locus coeruleus neurons in behaving rats anticipates fluctuations in the sleep–waking cycle. *Journal of Neuroscience*, **1**, 876–886.

Ax, A.F. (1953) The physiological differentiation between fear and anger in humans. *Psychosomatic Medicine*, **15**, 433–442.

Baker, R. and Bellis, M. (1995) *Human Sperm Competition: Copulation, Masturbation and Infidelity*. London: Chapman and Hall.

Bard, P.A. (1928) A diencephalic mechanism for the expression of rage with special reference to the sympathetic nervous system. *American Journal of Physiology*, **84**, 490–515.

Bartness, T.J., Powers, J.B., Hastings, M.H., Bittman, E.L. and Goldman, B.D. (1993) The timed infusion paradigm for melatonin delivery: what has it taught us about the melatonin signal, its reception, and the photoperiodic control of seasonal response? *Journal of Pineal Research*, **15**, 161–190.

Bateson, G., Jackson, D., Haley, J. and Weakland, J. (1956) Toward a theory of schizophrenia. *Behavioral Science*, **1(4)**, 251–264.

Beauchamp, G.K. and Cowart, B.J. (1993) Preferences for high salt concentration among children. *Developmental Psychology*, **26**, 539–545.

Ben-Dor, D.H., Laufer, N., Apter, A., Frisch, A. and Weizman, A. (2002) Heritability, genetics and association findings in anorexia nervosa. *Israeli Journal of Psychiatry and Related Sciences*, **39**, 262–270.

Benson, D.F. and Geschwind, N. (1985) The aphasias and related disturbances. In A.B. Baker and R.J. Joynt (eds) *Clinical Neurology*. New York: Harper and Row.

Berger, R.J. and Oswald, I. (1962) Effects of sleep deprivation on behaviour, subsequent sleep, and dreaming. *Journal of Mental Science*, **106**, 457–465.

Bermond, B., Nieuwenhuyse, B., Fasotti, L. and Scuerman, J. (1991) Spinal cord lesions, peripheral feedback, and intensities of emotional feelings. *Cognition and Emotion*, **5**, 201–220.

Bermudezrattoni, F. and McGaugh, J.L. (1991) Insular cortex and amygdala lesions differentially affect acquisition of inhibitory avoidance and conditioned taste aversion. *Brain Research*, **549**, 165–170.

Binder, J.R. (1997) Neuroanatomy of language processing studied with functional MRI. *Clinical Neuroscience*, **4**, 87–94.

Birdwhistell, R.L. (1970) *Kinetics and Context*. Philadelphia, PA: University of Pennsylvania Press.

Bliss, T.V.P. and Lømø, T. (1973) Long-lasting potentiation of synaptic transmission in the dentate area of the anaesthetized rabbit following stimulation of the perforant path. *Journal of Physiology*, **232**, 331–356.

Bloch, G.J. and Gorski, R.A. (1988) Cytoarchitectonic analysis of the SDN-POA of the intact and gonadectomised rat. *Journal of Comparative Neurology*, **275**, 604–612.

Booth, D.A. (1991) Influences on human food consumption. In D.J. Ramsay and D.A. Booth (eds) *Thirst: Physiological and Psychological Aspects*. London: Springer–Verlag.

Boulos, Z., Campbell, S.S., Lewy, A.J., Terman, M., Dijk, D.J. and Eastman, C.I. (1995) Light treatment for sleep disorders: consensus report. 7: jet-lag. *Journal of Biological Rhythms*, **10**, 167–176.

Brand, J.G. (2000) Receptor and transduction processes for umami taste. *Journal of Nutrition*, **130** (Supplement), 942S–945S.

Breggin, P.R. (1964) The psychophysiology of anxiety,

with a review of the literature concerning adrenaline. *Journal of Nervous and Mental Disease*, **139**, 558–568.

Breiter, H.C. et al. (1996) Functional magnetic resonance imaging of symptom provocation in obsessive–compulsive disorder. *Archives of General Psychiatry*, **53**, 595–606.

Brémer, F. (1936) Nouvelles recherches sur le mécanisme du sommeil. *Comtes Rendus de la Société de Biologie*, **22**, 460–464.

Bridges, R.S., Numan, M., Ronsheim, P.M., Mann, P.E. and Lupini, C.E. (1990) Central prolactin infusions stimulate maternal behavior in steroid-treated nulliparous female rats. *Proceedings of the National Academy of Sciences*, **87**, 8003–8007.

Brown, G.W., Birley, J.L.T. and Wing, J.K. (1972) Influence of family life on the course of schizophrenic disorders: a replication. *British Journal of Psychiatry*, **121**, 241–258.

Brown, R. (1995) Muramyl peptides and the functions of sleep. *Behavioural Brain Research*, **69**, 85–90.

Brownell, K.D., Greenwood, M.R.C., Stellar, E. and Shrager, E.E. (1986) The effect of repeated cycles of weight loss and regain in rats. *Physiology and Behavior*, **38**, 459–464.

Brunner, D.P., Dijk, D.J., Tobler, I. and Borbely, A.A. (1990) Effect of partial sleep deprivation on sleep stages and EEG power spectra: evidence for non-REM and REM sleep homeostasis. *Electroencephalography and Clinical Neurophysiology*, **75**, 492–499.

Bulik, C.M. et al. (2003) Significant linkage on chromosome 10p in families with bulimia nervosa. *American Journal of Human Genetics*, **72**, 200–207.

Butini, S. et al. (2003) Novel antipsychotic agents: recent advances in the drug treatment of schizophrenia. *Expert Opinion on Therapeutic Patents*, **13**, 425–448.

Cahill, L., Babinsky, R., Markowitsch, H.J. and McGaugh, J.L. (1995) The amygdala and emotional memory. *Nature*, **377**, 295–296.

Cami, J. et al. (2000) Human pharmacology of 3,4-methylenedioxymethamphetamine ('ecstasy'): psychomotor performance and subjective effects. *Journal of Clinical Psychopharmacology*, **20**, 455–466.

Campbell, S.S. and Tobler, I. (1985) Animal sleep: a review of sleep duration across phylogeny. *Neuroscience and Biobehavioral Reviews*, **8**, 269–300.

Cannon, W.B. (1927) The James–Lange theory of emotion: a critical examination and an alternative theory. *American Journal of Psychology*, **39**, 106–124.

Cannon, W.B. (1929) *Bodily Changes in Pain, Hunger, Fear, and Rage*. New York: Ronald.

Cannon, W.B. (1932) *The Wisdom of the Body*. New York: Norton.

Cannon, W.B. and Washburn, A.L. (1912) An explanation of hunger. *American Journal of Physiology*, **29**, 441–454.

Cannon, W.B., Newton, H.F., Bright, E.M., Menkin, V. and Moore, R.M. (1929) Some aspects of the physiology of animals surviving complete exclusion of sympathetic nerve impulses. *American Journal of Physiology*, **89**, 84–107.

Cartwright, R.D. (1979) The nature and function of repetitive dreams: a survey and speculation. *Psychiatry*, **42**, 131–137.

Caulliez, R., Meile, M.J. and Nicolaidis, S. (1996) A lateral hypothalamic Di dopaminergic mechanism in conditioned taste aversion. *Brain Research*, **729**, 234–245.

Chapman, I.M., Goble, E.A., Wittert, G.A., Morley, J.E. and Horowitz, M. (1998) Effect of intravenous glucose and exogenous insulin infusions on short-term appetite and food intake. *American Journal of Physiology*, **274**, 596–603.

Clark, J.T., Kalra, P.S., Crowley, W.R. and Kalra, S.P. (1984) Neuropeptide Y and human pancreatic polypeptide stimulate feeding behavior in rats. *Endocrinology*, **115**, 427–429.

Cohen, S. (1960) Lysergic acid diethylamide: side effects and complications. *Journal of Nervous and Mental Diseases*, **130**, 30–40.

Cole, J.L., Berman, N. and Bodner, R.J. (1997) Evaluation of chronic opioid receptor antagonist effects upon weight and intake measures in lean and obese Zucker rats. *Peptides*, **18**, 1201–1207.

Comuzzie, A.G. (2002) The emerging pattern of the genetic contribution to human obesity. *Best Practice and Research in Clinical Endocrinology and Metabolism*, **16**, 611–621.

Considine, R.V. et al. (1996) Serum immunoreactive leptin concentrations in normal-weight and obese humans. *New England Journal of Medicine*, **334**, 292–295.

Coren, S. (1996) *Sleep Thieves: An Eye-opening Exploration into the Science and Mysteries of Sleep*. New York: Free Press Paperbacks.

Crow, T.J. (1982) Positive and negative symptoms and the role of dopamine in schizophrenia. In G. Hemmings (ed) *Biological Aspects of Schizophrenia and Addiction*. New York: Wiley.

Damasio, A.R. (1996) The somatic marker hypothesis and the possible functions of the prefrontal cortex. *Philosophical Transactions of the Royal Society of London. Series B: Biological Sciences*, **351**, 1413–1420.

Dana, C.L. (1921) The anatomic seat of the emotions: a discussion of the James–Lange theory. *Archives of Neurology and Psychiatry*, **6**, 634.

Darwin, C. (1859) *The Origin of Species by Means of Natural Selection*. London: John Murray. (Harmondsworth: Penguin, 1968.)

Darwin, C. (1872) *The Expression of the Emotions in Man and Animals*. London: John Murray. (Definitive edition, P. Ekman (ed), 1998, London: HarperCollins.)

Davidson, R.J. (1992) Anterior cerebral asymmetry and the nature of emotion. *Brain and Cognition*, **20**, 125–151.

Davidson, R.J. et al. (1987) Ratings of emotion of faces are influenced by the visual field to which stimuli are presented. *Brain and Cognition*, **6**, 403–411.

Davis, J.D. and Campbell, C.S. (1973) Peripheral control of meal size in the rat: effect of sham feeding on meal size and drinking rate. *Journal of Comparative and Physiological Psychology*, **83**, 379–387.

Davis, K.L., Kahn, R.S., Ko, G. and Davidson, M. (1991). Dopamine in schizophrenia: a review and reconceptualization. *American Journal of Psychiatry*, **148**, 1471–1486.

Davis, M. (1992) The role of the amygdala in fear and anxiety. *Annual Review of Neuroscience*, **15**, 353–375.

Dawkins, R. (1989) *The Selfish Gene*, 2nd edn. Oxford: Oxford University Press.

Deacon, S. and Arendt, J. (1996) Adapting to phase

shifts. I. An experimental model for jet lag and shift work. *Physiology and Behavior*, **59**, 665–673.

Dement, W.C. (1978) *Some Must Watch while Some Must Sleep*. New York: W.W. Norton.

Dement, W.C. and Kleitman, N. (1957) The relation of eye movement during sleep to dream activity: an objective method for the study of dreaming. *Journal of Experimental Psychology*, **53**, 339–346.

Deutsch, J.A. and Gonzalez, M.F. (1980) Gastric nutrient content signals satiety. *Behavioral and Neural Biology*, **30**, 113–116.

Dittman, R.W., Kappes, M.E. and Kappes, M.H. (1992) Sexual behavior in adolescent and adult females with congenital adrenal hyperplasia. *Psychoneuroendocrinology*, **17**, 153–170.

Dorner, G., Schenck, B., Schmiedel, B. and Ahrens, L. (1983) Stressful events in prenatal life of bi- and homosexual men. *Experimental and Clinical Endocrinology*, **81**, 83–87.

Drevets, W.C. (1999) Prefrontal cortical–amygdalar metabolism in major depression. *Annals of the New York Academy of Sciences*, **877**, 614–637.

Drewnowski, A., Krahn, D.D., Demitrack, M.A., Nairn, K. and Gosnell, B.A. (1995) Naloxone, an opiate blocker, reduces the consumption of sweet high-fat foods in obese and lean female binge eaters. *American Journal of Clinical Nutrition*, **61**, 1206–1212.

Dronkers, N.F. (1996) A new brain region for coordinating speech articulation. *Nature*, **384**, 159–161.

Eagly, A.H. and Steffen, V.J. (1986) Gender and aggressive behavior: a meta-analytic review of the social psychological literature. *Psychological Bulletin*, **100**, 309–330.

Eden, G.F. et al. (1996) Abnormal processing of visual motion in dyslexia revealed by functional brain imaging. *Nature*, **382**, 66–69.

Ehrenkranz, J., Bliss, E. and Sheard, M. (1974) Plasma testosterone: correlation with aggressive behavior and social dominance in man. *Psychosomatic Medicine*, **36**, 469–475.

Ekman, P. (1992) An argument for basic emotions. *Cognition and Emotion*, **6**, 169–200.

Ekman, P. and Friesen, W.V. (1969) The repertoire of nonverbal behavior: categories, origins, usage, and coding. *Semiotica*, **1**, 49–98.

Engell, D. and Hirsch, E. (1991) Environmental and sensory modulation of fluid intake in humans. In D.J. Ramsay and D.A. Booth (eds) *Thirst: Physiological and Psychological Aspects*. London: Springer–Verlag.

Etcoff, N.L. (1984) Selective attention to facial identity and facial emotion. *Neuropsychologia*, **22**, 281–295.

Etcoff, N.L. (1989) Asymmetries in recognition of emotion. In F. Boller and J. Grafman (eds) *Handbook of Neuropsychology*. New York: Elsevier.

Evarts, E.V. (1974) Sensorimotor cortex activity associated with movements triggered by visual as compared to somesthetic inputs. In F.O. Schmitt and F.G. Worden (eds) *The Neurosciences*. Cambridge, MA: MIT Press.

Ferini-Strambi, L. and Zucconi, M. (2000) REM sleep behavior disorder. *Clinical Neurophysiology*, **111**, S136–S140.

Fitzsimmons, J.T. (1961) Drinking by rats depleted of body fluid without increase in osmotic pressure. *Journal of Physiology*, **159**, 297–309.

Fitzsimmons, J.T. and Moore-Gillon, M.J. (1980) Drinking and antidiuresis in response to reductions in venous return in the dog: neural and endocrine mechanisms. *Journal of Physiology*, **308**, 403–416.

Fleming, A. and Rosenblatt, J.S. (1974) Olfactory regulation of maternal behavior in rats. II. Effects of peripherally induced anosmia and lesions of the lateral olfactory tract in pup-induced virgins. *Journal of Comparative and Physiological Psychology*, **86**, 233–246.

Fleming, A., Vaccarino, F. and Luebke, C. (1980) Amygdaloid inhibition of maternal behavior in the nulliparous female rat. *Physiology and Behavior*, **25**, 731–745.

Fleming, A.S., Cheung, U., Myhal, N. and 'Kessler, Z. (1989) Effects of maternal hormones on 'timidity' and attraction to pup-related odors in female rats. *Physiology and Behavior*, **46**, 449–453.

Fleming, A.S., Steiner, M. and Corter, C. (1997) Cortisol, hedonics, and maternal responsiveness in human mothers. *Hormones and Behavior*, **32**, 85–98.

Floody, O.R. and Pfaff, D.W. (1977) Aggressive behavior in female hamsters: the hormonal basis for fluctuations in female aggressiveness correlated with estrus state. *Journal of Comparative and Physiological Psychology*, **91**, 443–464.

Francis, S. et al. (1999) The representation of pleasant touch in the brain and its relationship with taste and olfactory areas. *Neuroreport*, **10**, 453–459.

Fridlund, A. (1994) *Human Facial Expression: An Evolutionary View*. San Diego, CA: Academic Press.

Friedman, M. and Rosenman, R.H. (1959) Association of a specific overt behavior pattern with blood and cardiovascular findings. *Journal of the American Medical Association*, **169**, 1286–1296.

Frodl, T. et al. (2002) Hippocampal changes in patients with a first episode of major depression. *American Journal of Psychiatry*, **159**, 1112–1118.

Galaburda, A.M. et al. (1985) Developmental dyslexia: 4 consecutive patients with cortical anomalies. *Annals of Neurology*, **18**, 222–233.

Garcia, J. and Koelling, R.A. (1966) Relation of cue to consequence in avoidance learning. *Psychonomic Science*, **4**, 123–124.

Garcia, J., Hankins, W.G. and Rusiniak, K.W. (1974) Behavioral regulation of the melieu interne in man and rat. *Science*, **185**, 824–831.

Gazzaniga, M.S. and Sperry, R.W. (1967) Language after section of the cerebral commissure. *Brain*, **90**, 131–148.

Gazzaniga, M.S., Bogen, J.E. and Sperry, R.W. (1962) Some functional effects of sectioning the cerebral commissures in man. *Proceedings of the National Academy of Sciences*, **48**, 1765–1769.

George, M.S. et al. (1996) Understanding emotional prosody activates right hemisphere regions. *Archives of Neurology*, **53**, 665–670.

Gerkema, M.P. and Daan, S. (1985) Ultradian rhythms in behavior: the case of the common vole (*Microtus arvalis*). In H. Schultz and P. Lavie (eds) *Ultradian Rhythms in Physiology and Behavior*. London: Springer–Verlag.

Geschwind, N. (1965) Disconnexion syndromes in animals and man. *Brain*, **88**, 237–294.

Goldstein, K. (1952) The effect of brain damage on the personality. *Psychiatry*, **15**, 41–45.

Gonzalez, M.F. and Deutsch, J.A. (1981) Vagotomy abolishes cues of satiety produced by gastric distension. *Science*, **212**, 1283–1284.

Gonzalez-Mariscal, G. (2001) Neuroendocrinology of maternal behavior in the rabbit. *Hormones and Behavior*, **40**, 125–132.

Goodglass, H. (1993) *Understanding Aphasia*. New York: Academic Press.

Gorski, R.A., Gordon, J.H., Shryne, J.E. and Southam, A.M. (1978) Evidence for a morphological sex difference within the medial preoptic area of the rat brain. *Brain Research*, **148**, 333–346.

Gottesman, I.I. (1991) *Schizophrenia Genesis*. New York: W.H. Freeman.

Gregg, T.R. and Siegel, A. (2001) Brain structures and neurotransmitters regulating aggression in cats: implications for human aggression. *Progress in Neuro-Psychopharmacology and Biological Psychiatry*, **25**, 91–140.

Grill, H.J. and Kaplan, J.M. (2002) The neuroanatomical axis for control of energy balance. *Frontiers in Neuroendocrinology*, **23**, 2–40.

Grunt, J.A. and Young, W.C. (1953) Consistency of sexual behavior patterns in individual male guinea pigs following castration and androgen therapy. *Journal of Comparative and Physiological Psychology*, **46**, 138–144.

Guan, D., Phillips, W. and Green, G. (1996) Inhibition of gastric emptying and food intake by exogenous and endogenous CCK: role of capsaicin-sensitive vagal afferent pathway. *Gastroenterology*, **110**, A672.

Gurevich, E.V. et al. (1997) Mesolimbic dopamine D-3 receptors and use of antipsychotics in patients with schizophrenia — a postmortem study. *Archives of General Psychiatry*, **54**, 225–232.

Gurvits, T.V. et al. (1996) Magnetic resonance imaging study of hippocampal volume in chronic, combat-related posttraumatic stress disorder. *Biological Psychiatry*, **40**, 1091–1099.

Hall, W., Solowij, N. and Lemon, J. (1994) The health and psychological consequences of cannabis use. *National Drug Strategy Monograph Series*, No. 25. Canberra: Australian Government Publishing Service.

Harris, G. and Booth, D.A. (1987) Infants' preference for salt in food: its dependence upon recent dietary experience. *Journal of Reproductive and Infant Psychology*, **5**, 97–104.

Harris, L.J. (1978) Sex differences in spatial ability: possible environmental, genetic, and neurological factors. In M. Kinsbourne (ed) *Asymmetrical Function of the Human Brain*. Cambridge: Cambridge University Press.

Hatzinger, M. et al. (2002) The combined DEX-CRH test in treatment course and long-term outcome of major depression. *Journal of Psychiatric Research*, **36**, 287–297.

Hebb, D.O. (1949) *Organization of Behavior*. New York: Wiley.

Heimer, I. and Larsson, K. (1966) Impairment of mating behavior in male rats following lesions of the preoptic–anterior hypothalamic continuum. *Brain Research*, **3**, 248–263.

Heller, H.C. and Glotzbach, S.F. (1985) Thermoregulation and sleep. In E.C. Eberhardt and A. Shitzer (eds) *Heat Transfer in Biological Systems: Analysis and Application*. New York: Plenum.

Hennessey, A.C., Camak, L., Gordon, F. and Edwards, D.A. (1990) Connections between the pontine central grey and the ventromedial hypothalamus are essential for lordosis in female rats. *Behavioral Neuroscience*, **104**, 477–488.

Hetherington, M.H. and Rolls, B.J. (1996) Sensory-specific satiety: theoretical frameworks and central characteristics. In E.D. Capaldi (ed) *Why We Eat What We Eat*. Washington, DC: American Psychological Association.

Hetherington, M.M., Pirie, L.M. and Nabb, S. (2002) Stimulus satiation: effects of repeated exposure to foods on pleasantness and intake. *Appetite*, **38**, 19–28.

Hinton, G.E., Plaut, D.C. and Shallice, T. (1993) Simulating brain damage. *Scientific American*, **269**, 276–282.

Hoaken, P.N.S., Giancola, P.R. and Pihl, R.O. (1998) Executive cognitive functions as mediators of alcohol-related aggression. *Alcohol and Alcoholism*, **33**, 47–54.

Hohmann, G.W. (1966) Some effects of spinal cord lesions on experienced emotional feelings. *Psychophysiology*, **3**, 143–156.

Holmes, T.H. and Rahe, R.H. (1967) The Social Readjustment Rating Scale. *Journal of Psychosomatic Research*, **11**, 213–218.

Hopf, H.C., Mueller-Forell, W. and Hopf, N.J. (1992) Localization of emotional and volitional facial paresis. *Neurology*, **42**, 1918–1923.

Horne, J.A. (1978) A review of the biological effects of total sleep deprivation in man. *Biological Psychology*, **7**, 55–102.

Horne, J.A. (1988) *Why We Sleep. The Functions of Sleep in Humans and Other Mammals*. Oxford: Oxford University Press.

Hubel, D.H. and Wiesel, T.N. (1979) Brain mechanisms of vision. *Scientific American*, **249**, 150–162.

Hyde, J.S. (1986) Gender differences in aggression. In J.S. Hyde and M.C. Linn (eds) *The Psychology of Gender*. Baltimore, MD: Johns Hopkins University Press.

Isenberg, N. et al. (1999) Linguistic threat activates the human amygdala. *Proceedings of the National Academy of Sciences*, **96**, 10456–10459.

Jacobsen, C.F., Wolfe, J.B. and Jackson, T.A. (1935) An experimental analysis of the functions of the frontal association areas in primates. *Journal of Nervous and Mental Disease*, **82**, 1–14.

James, W. (1884) What is an emotion? *Mind*, **9**, 188–205.

Jänig, W. (2003) The autonomic nervous system and its coordination by the brain. In R.J. Davidson, K.R. Scherer and H.H. Goldsmith (eds) *Handbook of Affective Sciences*. Oxford/New York: Oxford University Press.

Jenkins, I.H., Brooks, D.J., Bixon, P.D., Frackowiak, R.S. and Passingham, R.E. (1994) Motor sequence learning: a study with positron emission tomography. *Journal of Neuroscience*, **14**, 3775–3790.

John, E.R. (1972) Switchboard versus statistical theories of learning and memory. *Science*, **177**, 850–864.

Kaas, J.H., Nelson, R.J., Sur, M. and Merzenich, M.M. (1981) Organization of somatosensory cortex in primates. In F.O. Schmitt, F.G. Worden, G. Adelman and S.G. Dennis (eds) *The Organization of the Cerebral Cortex*. Cambridge, MA: MIT Press.

Kandel, E.R. and Spencer, W.A. (1968) Cellular neurophysiological approaches in the study of learning. *Physiological Review*, **48**, 65–134.

Kandel, E.R., Hawkins, R.D., Antonov, I. and Antonova, I. (2001) The contribution of activity-dependent synaptic plasticity to classical conditioning in aplysia. *Journal of Neuroscience*, **21**, 6413–6422.

Keillor, J.M., Barrett, A.M., Crucian, G.P., Kortenkamp, S. and Heilman, K.M. (2002) Emotional experience and perception in the absence of facial feedback. *Journal of the International Neuropsychological Society*, **8**, 130–135.

Keith, S.J., Regier, D.A. and Rae, D.S. (1991) Schizophrenic disorders. In A.J. Frances and R.E. Hales (eds) *Review of Psychiatry*, vol. 7. Washington, DC: American Psychiatric Press.

Kelsoe, J.R. (2003) Arguments for the genetic basis of the bipolar spectrum. *Journal of Affective Disorders*, **73**, 183–197.

Kendell, R.E. and Adams, W. (1991) Unexplained fluctuations in the risk for schizophrenia by month and year of birth. *British Journal of Psychiatry*, **158**, 758–763.

Kertesz, A. (1981) Anatomy of jargon. In J. Brown (ed) *Jarganophasia*. New York: Academic Press.

Kertesz, A., Sheppard, A. and Mackenzie, R. (1982) Localization in transcortical sensory aphasia. *Archives of Neurology*, **39**, 475.

Kevenau, J.L. (1997) Origin and evolution of sleep: roles of vision and endothermy. *Brain Research Bulletin*, **42**, 245–264.

Kish, S.J. (2002) How strong is the evidence that brain serotonin neurons are damaged in human users of ecstasy? *Pharmacology, Biochemistry and Behavior*, **71**, 845–855.

Kleitman, N. (1961) The nature of dreaming. In G.E.W. Wolstenholme and M. O'Connor (eds) *The Nature of Sleep*. London: J&A Churchill.

Klüver, H. and Bucy, P.C. (1938) An analysis of certain effects of bilateral temporal lobectomy in the rhesus monkey, with special reference to 'psychic blindness'. *Journal of Psychology*, **5**, 33–54.

Kondziolka, D. (1999) Functional radiosurgery. *Neurosurgery*, **44**, 12–20.

Koopmans, H.S. (1981) The role of the gastrointestinal tract in the satiation of hunger. In L.A. Cioffi, W.B.T. James and T.B. VanItalie (eds) *The Body Weight Regulatory System: Normal and Disturbed Mechanisms*. New York: Raven Press.

Kotz, C.M., Briggs, J.E., Pomonis, J.D., Grace, M.K., Levine, A.S. and Billington, C.J. (1998) Neural site of leptin influence on neuropeptide Y signaling pathways altering feeding and uncoupling protein. *American Journal of Physiology*, **275**, R478–R484.

Kraly, F.S. (1990) Drinking elicited by eating. In A.N. Epstein and A. Morrison (eds) *Progress in Psychobiology and Physiological Psychology*. New York: Academic Press.

Krystal, J.H. et al. (1992) Chronic 3,4-methylenedioxymethamphetamine (MDMA) use: effects on mood and neuropsychological function. *American Journal of Drug and Alcohol Abuse*, **18**, 331–341.

Laird, J.D. (1974) Self-attribution of emotion: the effects of expressive behavior on the quality of emotional experience. *Journal of Personality and Social Psychology*, **29**, 475–486.

Lashley, K.S. (1929) *Brain Mechanisms and Intelligence*. Chicago, IL: University of Chicago Press.

Lavond, D.G., Knowlton, B.J., Steinmetz, J.E. and Thompson, R.F. (1987) Classical conditioning of the rabbit eyelid response with a mossy-fiber stimulation CS: II. Lateral reticular nucleus stimulation. *Behavioral Neuroscience*, **101**, 676–682.

LeDoux, J.E. (1995) Emotion: clues from the brain. *Annual Review of Psychology*, **46**, 209–235.

Le Magnen, J. (1990) A role for opiates in food reward and food addiction. In E.D. Capaldi and T.L. Powley (eds) *Taste, Experience and Feeding*. Washington, DC: American Psychological Association.

Leonard, B.E. (2000) Evidence for a biochemical lesion in depression. *Journal of Clinical Psychiatry*, **61**, 12–17.

Leroi I., O'Hearn E., Marsh L. et al. (2002) Psychopathology in patients with degenerative cerebellar diseases: a comparison to Huntington's disease. *American Journal of Psychiatry*, **159**, 1306–1314.

LeVay, S. (1991) A difference in hypothalamic structure between heterosexual and homosexual men. *Science*, **253**, 1034–1037.

Lind, R.W. and Johnson, A.K. (1982) Central and peripheral mechanisms mediating angiotensin-induced thirst. In D. Ganten, M. Printz and B.A. Schölkens (eds) *The Renin Angiotensin System in the Brain*. London: Springer–Verlag.

Lisk, R.D., Pretlow, R.A. and Friedman, S. (1969) Hormonal stimulation necessary for elicitation of maternal nest-building in the mouse (*Mus musculus*). *Animal Behavior*, **17**, 730–737.

Lonstein, J.S. and Gammie, S.C. (2002) Sensory, hormonal, and neural control of maternal aggression in laboratory rodents. *Neuroscience and Biobehavioral Reviews*, **26**, 869–888.

Lopez-Garcia, J.A. (2003) Spinal memory of peripheral inflammation. *Journal of Physiology*, **548**, 1S.

Lowe, J. and Carroll, D. (1985) The effects of spinal cord injury on the intensity of emotional experience. *British Journal of Clinical Psychology*, **24**, 135–136.

Lyamin, O.I. (1993) Sleep in the harp seal (*Pagophilus groenlandica*): comparison of sleep on land and in water. *Journal of Sleep Research*, **2**, 170–174.

MacKinnon, D.F. et al. (2002) Comorbid bipolar disorder and panic disorder in families with a high prevalence of bipolar disorder. *American Journal of Psychiatry*, **159**, 30–35.

Mahowald, M.W., Bundlie, S.R., Hurwitz, T.D. and Schenck, C.H. (1990) Sleep violence — forensic science implications: polygraphic and video documentation. *Journal of Forensic Science*, **35**, 413–432.

Malsbury, C.W. (1972) Facilitation of male rat copulatory behavior by electrical stimulation of the medial preoptic area. *Physiology and Behavior*, **7**, 797–805.

Mann, M.A., Konen, C. and Svare, B. (1984) The role of progesterone in pregnancy-induced aggression in mice. *Hormones and Behavior*, **18**, 140–160.

Manstead, A.S.R. and Wagner, H.L. (1981) Arousal, cognition and emotion: an appraisal of two-factor theory. *Current Psychological Reviews*, **1**, 35–54.

Maquet, P. (2001) The role of sleep in learning and memory. *Science*, **294**, 1048–1052.

Marañon, G. (1924) Contribution à l'étude de l'action émotive de l'adrenaline. *Revue Française d'Endocrinologie*, **2**, 301–325.

Martin, R.J., White, B.D. and Hulsey, M.G. (1991) The regulation of body weight. *American Scientist*, **79**, 528–541.

Marx, R.D. (1994) Anorexia nervosa: theories of etiology. In L. Alexander-Mott and D.B. Lumsden (eds) *Understanding Eating Disorders: Anorexia Nervosa, Bulimia Nervosa, and Obesity*. Philadelphia, PA: Taylor and Francis.

Mas, M. (1995) Neurobiological correlates of masculine sexual behavior. *Neuroscience and Biobehavioral Reviews*, **19**, 261–277.

Mayberg, H.S. et al. (1999) Reciprocal limbic–cortical function and negative mood: converging PET findings in depression and normal sadness. *American Journal of Psychiatry*, **156**, 675–682.

Mayer, J. (1953) Glucostatic mechanisms of regulation of food intake. *New England Journal of Medicine*, **249**, 13–16.

Mazur, A. and Lamb, T. (1980) Testosterone, status, and mood in human males. *Hormones and Behavior*, **14**, 136–146.

Mbugua, K. (2003) Sexual orientation and brain structures: a critical review of recent research. *Current Science*, **84**, 173–178.

McEwen, B.S. and Seeman, T. (2003) Stress and affect: applicability of the concepts of allostasis and allostatic load. In R.J. Davidson, K.R. Scherer and H.H. Goldsmith (eds) *Handbook of Affective Sciences*. Oxford/New York: Oxford University Press.

McEwen, B.S. et al. (1997) Neural–endocrine–immune interactions: the role of adrenocorticoids as modulators of immune function. *Brain Research Reviews*, **23**, 79–133.

McGinty, D.J. and Sterman, M.B. (1968) Sleep suppression after basal forebrain lesions in the cat. *Science*, **160**, 1253–1255.

McGinty, D.J., Szymusiak, R. and Thompson, D. (1994) Preoptic/anterior hypothalamic warming increases EEG delta frequency activity within non-REM sleep. *Brain Research*, **667**, 273–277.

McIntosh, J.L. (1991) Epidemiology of suicide in the U.S. In A.A. Leenaars (ed) *Life Span Perspectives of Suicide*. New York: Plenum Press.

McKinley, M.J., Pennington, G.L. and Oldfield, B.J. (1996) Anteroventral wall of the third ventricle and dorsal lamina terminalis: headquarters for control of body fluid homeostasis? *Clinical and Experimental Pharmacology and Physiology*, **23**, 271–281.

McKinney, T.D. and Desjardins, C. (1973) Postnatal development of the testis, fighting behavior, and fertility in house mice. *Biology of Reproduction*, **9**, 279–294.

McNeill, E.T. (1994) Blood, sex and hormones: a theoretical review of women's sexuality over the menstrual cycle. In P.Y.L. Choi and P. Nicholson (eds) *Female Sexuality: Psychology, Biology and Social Context*. Hemel Hempstead: Harvester.

Melzack, R. and Wall, P.D. (1965) Pain mechanisms: a new theory. *Science*, **150**, 971–979.

Mennella, J.A. and Beauchamp, G.K. (1996) The early development of human flavor preferences. In E.D. Capaldi (ed) *Why We Eat What We Eat*. Washington, DC: American Psychological Association.

Meyer-Bahlburg, H.F.L. (1984) Psychoendocrine research on sexual orientation: current status and future options. *Progress in Brain Research*, **63**, 375–398.

Milner, B. (1974) Hemispheric specialization: Scope and limits. In F.O. Schmitt and F.G. Worden (eds) *The Neurosciences: Third Study Program*. Cambridge, MA: MIT Press.

Mistleberger, R.E. (1994) Circadian food-anticipatory activity: formal models and physiological mechanisms. *Neuroscience and Biobehavioral Reviews*, **18**, 171–195.

Money, J. and Ehrhardt, A. (1972) *Man and Woman, Boy and Girl*. Baltimore, MD: Johns Hopkins University Press.

Monroe, R.R. (1978) *Brain Dysfunction in Aggressive Criminals*. Lexington, MA: Lexington Books.

Moore, R.Y. and Eichler, V.B. (1972) Loss of a circadian adrenal corticosterone rhythm following suprachiasmatic lesions in the rat. *Brain Research*, **42**, 201–206.

Moore-Ede, M.C. (1992) *The Clocks that Time Us: Physiology of the Circadian Timing System*. Cambridge, MA: Harvard University Press.

Morgan, C.T. and Morgan, J.D. (1943) Studies in hunger: II. The relation of gastric denervation and dietary sugar to the effect of insulin upon food hoarding in the rat. *Journal of Genetic Psychology*, **57**, 153–163.

Morris, D.W. et al. (2003) No evidence for association of the dysbindin gene (DTNBP1) with schizophrenia in an Irish population-based study. *Schizophrenia Research*, **60**, 167–172.

Morris, G.O., Williams, H.L. and Lubin, A. (1960) Misperception and disorientation during sleep deprivation. *Archives of General Psychiatry*, **2**, 247–254.

Morris, R.G.M., Garrud, P., Rawlins, J.N.P. and O'Keefe, J. (1982) Place navigation impaired in rats with hippocampal lesions. *Nature*, **297**, 681–683.

Moruzzi, G. and Magoun, H.W. (1949) Brain stem reticular formation and activation of the EEG. *Electroencephalography and Clinical Neurophysiology*, **1**, 455–473.

Mukhamentov, L.M., Supin, A.Y. and Lyamin, O.I. (1988) Interhemispheric asymmetry of the EEG during sleep in marine mammals. In T. Oniani (ed) *The Neurobiology of the Sleep–Wakefulness Cycle*. Tbilisi: Montenierebe.

Naeser, M.A., Hayward, R.W., Laughlin, S.A. and Zatz, L.M. (1981) Quantitative CT scan studies in aphasia. *Brain and Language*, **12**, 140–164.

Naranjo, C.A., Tremblay, L.K. and Busto, U.E. (2001) The role of the brain reward system in depression. *Progress in Neuro-psychopharmacology and Biological Psychiatry*, **25**, 781–823.

Nemeroff, C.B. (1998) The neurobiology of depression. *Scientific American*, **278**, 42–29.

Nicolson, R.I., Fawcett, A.J. and Dean, P. (2001) Developmental dyslexia: the cerebellar deficit hypothesis. *Trends in Neurosciences*, **24**, 508–511.

Nishino, S., Reid, M.S. Dement, W.C. and Mignot, E. (1994) Neuropharmacology and neurochemistry of canine narcolepsy. *Sleep*, **17**, S84–S92.

Numan, M. (1974) Medial preoptic area and maternal behavior in the female rat. *Journal of Comparative and Physiological Psychology*, **87**, 746–759.

Nutt, D.J. (2002) The neuropharmacology of serotonin and noradrenaline in depression. *International Clinical Psychopharmacology*, **17** (Supplement), S1–S12.

Ohyama, T., Nores, W.L., Murphy, M., and Mauk, M.D. (2003) What the cerebellum computes. *Trends in Neurosciences*, **26**, 222–227.

O'Keefe, J. and Nadel, L. (1978) *The Hippocampus as a Cognitive Map*. Oxford: Oxford University Press.

Olds, J. (1962) Hypothalamic substrates of reward. *Physiological Reviews*, **42**, 554–604.

Olds, J. and Milner, P. (1954) Positive reinforcement produced by electrical stimulation of septal area and other regions of rat brain. *Journal of Comparative and Physiological Psychology*, **47**, 419–427.

Olton, D.S. (1983) Memory functions in the hippocampus. In W. Seifert (ed) *Neurobiology of the Hippocampus*. New York: Academic Press.

Overmier, J.B. and Murison, R. (1997) Animal models reveal the 'psych-' in the psychosomatics of peptic ulcers. *Current Directions in Psychological Science*, **6**, 180–184.

Papez, J.W. (1937) A proposed mechanism of emotion. *Archives of Neurology and Psychiatry*, **38**, 725–743.

Pavlov, I.P. (1927) *Conditioned Reflexes*. Oxford: Oxford University Press.

Pecina, S. and Berridge, K.C. (1996) Brainstem mediates diazepam enhancement of palatability and feeding: microinjections into fourth ventricle versus lateral ventricle. *Brain Research*, **727**, 22–30.

Pedrazzi, P., Cattaneo, L., Valeriani, L., Boschi, S., Cocchi, D. and Zoli, M. (1998) Hypothalamic neuropeptide Y and galanin in overweight rats fed a cafeteria diet. *Peptides*, **19**, 157–165.

Penfield, W. and Boldrey, E. (1937) Somatic motor and sensory representations in the cerebral cortex of man as studied by electrical stimulation. *Brain*, **60**, 389–443.

Penfield, W. and Perot, P. (1963) The brain's record of auditory and visual experience. *Brain*, **86**, 595–696.

Penfield, W. and Rasmussen, T. (1950) *The Cerebral Cortex of Man: a Clinical Study of Localization*. Boston: Little, Brown.

Penfield, W. and Roberts, L. (1959) *Speech and Brain Mechanisms*. Princeton, NJ: Princeton University Press.

Peroutka, S.J., Newman, H. and Harris, H. (1988) Subjective effects of 3,4-methylenedioxymethamphetamine (MDMA) in recreational users. *Neuropsychopharmacology*, **1**, 273–277.

Petursson, H. and Lader, M.H. (1981) Withdrawal from long term benzodiazepine treatment. *British Medical Journal*, **283**, 643–645.

Pfaff, D.W. and Sakuma, Y. (1979) Deficit in the lordosis reflex of female rats caused by lesions in the ventromedial nucleus of the hypothalamus. *Journal of Physiology*, **288**, 203–210.

Phoenix, C.H., Goy, R.W., Gerall, A.A. and Young, W.C. (1959) Organizing action of prenatally administered testosterone propionate on the tissues mediating mating behavior in the female guinea pig. *Endocrinology*, **65**, 369–382.

Piéron, H. (1913) *Le Problème Physiologique du Sommeil*. Paris: Masson.

Post, R.M., Ballenger, J.C. and Goodwin, F.K. (1980) Cerebrospinal fluid studies of neurotransmitter in manic and depressive illness. In J.H. Wood (ed) *The Neurobiology of Cerbrospinal Fluid*, vol. 1. New York: Plenum.

Pryce, C.R. (1993) The regulation of maternal behavior in marmosets and tamarins. *Behavioural Processes*, **30**, 201–224.

Quigg, M. and Fountain, N.B. (1999) Conduction aphasia elicited by stimulation of the left posterior superior temporal gyrus. *Journal of Neurology, Neurosurgery and Psychiatry*, **66**, 393–396.

Ramsay, D.J., Rolls, B.J. and Wood, R.J. (1977) Thirst following water deprivation in dogs. *American Journal of Physiology*, **232**, 93–100.

Rand, C.S.W. (1994) Obesity: definition, diagnostic criteria, and associated health problems. In L. Alexander-Mott and D.B. Lumsden (eds) *Understanding Eating Disorders: Anorexia Nervosa, Bulimia Nervosa, and Obesity*. Philadelphia, PA: Taylor and Francis.

Rattenborg, N.C., Lima, S.L. and Amlaner, C.J. (1999) Facultative control of avian unihemispheric sleep under the risk of predation. *Behavioural Brain Research*, **105**, 163–172.

Reason, J., Wagner, H. and Dewhurst, D. (1981) A visually-driven postural after-effect. *Acta Psychologica*, **48**, 241–251.

Rechtschaffen, A. and Bergmann, B.M. (1995) Sleep deprivation in the rat by the disk-over-water method. *Behavioural Brain Research*, **69**, 55–63.

Rhees, R.W., Shryne, J.E. and Gorski, R.A. (1990) Termination of the hormone-sensitive period for differentiation of the sexually dimorphic nucleus of the preoptic area in male and female rats. *Developmental Brain Research*, **52**, 17–23.

Rickels, K., Schweizer, E., Weiss, S. and Zavodnick, S. (1993) Maintenance drug treatment for panic disorder II. Short-term and long-term outcome after drug taper. *Archives of General Psychiatry*, **50**, 61–68.

Ridley, M. (1994) *The Red Queen: Sex and the Evolution of Human Nature*. Harmondsworth: Penguin Books.

Rinn, W.E. (1984) The neuropsychology of facial expression: a review of the neurological and psychological mechanisms for producing facial expressions. *Psychological Bulletin*, **95**, 52–77.

Robinson, T.E. and Berridge, K.C. (1993) The neural basis of drug craving: an incentive-sensitization theory of addiction. *Brain Research Review*, **18**, 247–291.

Roffwarg, H.P., Muzio, J.N. and Dement, W.C. (1966) Ontogenetic development of human sleep–dream cycle. *Science*, **152**, 604–619.

Rolls, E.T. (1982) Feeding and reward. In B.G. Hoebel and D. Novin (eds) *The Neural Basis of Feeding and Reward*. Brunswick, Maine: Haer Institute for Electrophysiological Research.

Rolls, E.T. (1993) The neural control of feeding in primates. In D.A. Booth (ed) *The Neurophysiology of Ingestion*. Elmsford, NY: Pergamon Press.

Rosenberger, P.B. (1990) Morphological cerebral asymmetries and dyslexia. In G. Th. Pavlidis (ed) *Perspectives on Dyslexia*, vol. 1. Chichester/New York: Wiley.

Rosenman, R.H. et al. (1964) A productive study of coronary heart disease: The Western Collaborative Group Study. *Journal of the American Medical Association*, **189**, 15–22.

Rothwell, N.J. and Stock, M.J. (1982) Energy expenditure derived from measurements of oxygen consumption and energy balance in hyperphagic, 'cafeteria'-fed rats. *Journal of Physiology*, **324**, 59–60.

Russell, M.J., Switz, G.M. and Thompson, K. (1980) Olfactory influences on the human menstrual cycle. *Pharmacology, Biochemistry, and Behavior*, **13**, 737–738.

Sato, W. et al. (2002) Seeing happy emotion in fearful and angry faces: qualitative analysis of facial expression recognition in a bilateral amygdala-damaged patient. *Cortex*, **38**, 727–742.

Saxena, S. et al. (1999) Localized orbitofrontal and subcortical metabolic changes and predictors of response to paroxetine treatment in obsessive–compulsive disorder. *Neuropsychopharmacology*, **21**, 683–693.

Schachter, S. (1964) The interaction of cognitive and physiological determinants of emotional state. In L. Berkowitz (ed) *Advances in Experimental Social Psychology*, vol. 1. New York: Academic Press.

Schachter, S. and Singer, J.E. (1962) Cognitive, social and physiological determinants of emotional state. *Psychological Review*, **69**, 379–399.

Schneider, F., Habel, U., Volkmann, J., Regel, S., Kornischka, J., Sturm, V. and Freund, H.J. (2003) Deep brain stimulation of the subthalamic nucleus enhances emotional processing in Parkinson disease. *Archives of General Psychiatry*, **60**, 296–302.

Schultz, W., Tremblay, L. and Hollerman, J.R. (1998) Reward prediction in primate basal ganglia and frontal cortex. *Neuropharmacology*, **37**, 421–429.

Schwartz, W.J. and Gainer, H. (1977) Suprachiasmatic nucleus: use of ^{14}C-labelled deoxyglucose uptake as a functional marker. *Science*, **197**, 1089–1091.

Sclafani, A. and Nissenbaum, J.W. (1988) Robust conditioned flavor preference produced by intragastric starch infusion in rats. *American Journal of Physiology*, **255**, R672–R675.

Scoville, W.B. and Milner, B. (1957) Loss of recent memory after bilateral hippocampal lesions. *Journal of Neurology, Neurosurgery and Psychiatry*, **20**, 11–21.

Segman, R.H. et al. (2002) Association between the dopamine transporter gene and posttraumatic stress disorder. *Molecular Psychiatry*, **7**, 903–907.

Seligman, M.E.P. (1971) Phobias and preparedness. *Behavior Therapy*, **2**, 307–320.

Selye, H. (1936) A syndrome produced by diverse noxious agents. *Nature*, **138**, 32.

Shabsigh, R. (1997) The effects of testosterone on the cavernous tissue and erectile function. *World Journal of Urology*, **15**, 21–26.

Shaldubina, A., Agam, G. and Belmaker, R.H. (2001) The mechanism of lithium action: state of the art 10 years later. *Progress in Neuro-psychopharmacology and Biological Psychiatry*, **25**, 855–866.

Sham, P.C. et al. (1992) Schizophrenia following prenatal exposure to influenza epidemics between 1939 and 1960. *British Journal of Psychiatry*, **160**, 461–466.

Shenton, M.E., Dickey, C.C., Frumin, M. and McCarley, R.W. (2001) A review of MRI findings in schizophrenia. *Schizophrenia Research*, **49**, 1–52.

Sherrington, C.S. (1900) Experiments on the value of vascular and visceral factors for the genesis of emotion. *Proceedings of the Royal Society*, **66**, 390–403.

Sherwin, B.B. (1988) A comparative analysis of the role of androgen in human male and female sexual behavior: behavioral specificity, critical thresholds, and sensitivity. *Psychobiology*, **16**, 416–425.

Sherwin, B.B. and Gelfand, M.M. (1987) The role of androgen in the maintenance of sexual functioning in oopherectomized women. *Psychosomatic Medicine*, **49**, 397–409.

Shouse, M.N. and Siegel, J.M. (1992) Pontine regulation of REM sleep components in cats: integrity of the pedunculopontine tegmentum is important for phasic events but unnecessary for atonia during REM sleep. *Brain Research*, **571**, 50–63.

Siegel, A., Roeling, T.A.P., Gregg, T.R. and Kruk, M.R. (1999) Neuropharmacology of brain-stimulation-evoked aggression. *Neuroscience and Biobehavioral Reviews*, **23**, 359–389.

Siegel, J.M. (1995) Phylogeny and the function of REM sleep. *Behavioural Brain Research*, **69**, 29–34.

Siegel, J.M. (2001) The REM sleep–memory consolidation hypothesis. *Science*, **294**, 1058–1063.

Sims, E.A.H. and Horton, E.S. (1968) Endocrine metabolic adaptation to obesity and starvation. *American Journal of Clinical Nutrition*, **21**, 1455–1470.

Singewald, N. and Sharp, T. (2000) Neuroanatomical targets of anxiogenic drugs in the hindbrain as revealed by Fos immunocytochemistry. *Neuroscience*, **98**, 759–770.

Skinner, B.F. (1938) *The Behavior of Organisms*. New York: Appleton–Century–Crofts.

Smith, G.P. and Gibbs, J. (1994) Satiating effect of cholecystokinin. *Annals of the New York Academy of Sciences*, **713**, 236–241.

Soderpalm, A.H.V. and Berridge, K.C. (2000) The hedonic impact and intake of food are increased by midazolam microinjection in the parabrachial nucleus. *Brain Research*, **877**, 288–297.

Springer, S.P. and Deutsch, G. (1997) *Left Brain, Right Brain: Perspectives from Cognitive Science*. New York: W.H. Freeman.

Stebbins, W.C., Miller, J.M., Johnsson, L.G. and Hawkins, J.E. (1969) Ototoxic hearing loss and cochlear pathology in the monkey. *Annals of Otology, Rhinology, and Laryngology*, **78**, 1007–1026.

Stein, M.B. et al. (2002) Genetic and environmental influences on trauma exposure and posttraumatic stress disorder symptoms: a twin study. *American Journal of Psychiatry*, **159**, 1675–1681.

Stephan, F.K. and Zucker, I. (1972) Circadian rhythms in drinking behavior and locomotor behavior of rats are eliminated by hypothalamic lesions. *Proceedings of the National Academy of Sciences*, **69**, 1583–1586.

Steriade, M., Paré, D., Datta, S., Oakson, G. and Curró Dossi, R. (1990) Different cellular types in mesopontine cholinergic nuclei related to PGO waves. *Journal of Neuroscience*, **8**, 2560–2579.

Stern, K. and McClintock, M.K. (1998) Regulation of ovulation by human pheromones. *Nature*, **392**, 177–179.

Straub, R.E. et al. (2002) Genetic variation of the 6p22.3 gene DTNBP1, the human ortholog of the mouse dysbindin gene, is associated with schizophrenia. *American Journal of Human Genetics*, **71**, 337–348.

Strecker, R.E. et al. (2000) Adenosinergic modulation of basal forebrain and preoptic/anterior hypothalamic neuronal activity in the control of behavioral state. *Behavioural Brain Research*, **115**, 183–204.

Stunkard, A.J. et al. (1986) An adoption study of human obesity. *New England Journal of Medicine*, **314**, 193–198.

Stuss, D.T., Guberman, A., Nelson, R. and Larochelle, S.

(1988) The neuropsychology of paramedian thalamic infarction. *Brain and Cognition*, **8**, 348–378

Sullivan, P.F., Neale, M.C. and Kendler, K.S. (2000) Genetic epidemiology of major depression: review and meta-analysis. *American Journal of Psychiatry*, **157**, 1552–1562.

Svare, B., Mann, M.A., Broida, J. and Michael, S. (1982) Maternal aggression exhibited by hypophysectomized parturient mice. *Hormones and Behavior*, **16**, 455–461.

Swaab, D.F. and Hofman, M.A. (1990) An enlarged suprachiasmatic nucleus in homosexual men. *Brain Research*, **537**, 141–148.

Swaab, D.F., Slob, A.K., Houtsmuller, E.J., Brand, T. and Zhou, J.N. (1995) Increased number of vasopressin neurons in the suprachiasmatic nucleus (SCN) of bisexual adult male rats following perinatal treatment with the aromatase blocker ATD. *Developmental Brain Research*, **85**, 273–279.

Swithers, S.E. and Hall, W.G. (1994) Does oral experience terminate ingestion? *Appetite*, **23**, 113–138.

Takahashi, L.K. (1990) Hormonal regulation of sociosexual behavior in female mammals. *Neuroscience and Biobehavioral Reviews*, **14**, 403–413.

Takami, S. (2002) Recent progress in the neurobiology of the vomeronasal organ. *Microscopy Research and Technique*, **58**, 228–250.

Tei, H., Okamura, H., Shigeyoshi, Y., Fukuhara, C., Ozawa, R., Hirose, M. and Sakaki, Y. (1997) Circadian oscillation of a mammalian homologue of the Drosophila period gene. *Nature*, **389**, 512–516.

Teitelbaum, P. and Stellar, E. (1954) Recovery from failure to eat produced by hypothalamic lesions. *Science*, **120**, 894–895.

Terman, M., Amira, L., Terman, J.S. and Ross, D.C. (1996) Predictors of response and nonresponse to light treatment for winter depression. *American Journal of Psychiatry*, **153**, 1423–1429.

Thut, G. et al. (1997) Activation of the human brain by monetary reward. *Neuroreport*, **8**, 1225–1228.

Timmann D., Drepper J., Maschke M., Kolb F.P., Boring D., Thilmann A.F. and Diener H.C. (2002) Motor deficits cannot explain impaired cognitive associative learning in cerebellar patients. *Neuropsychologia*, **40**, 788–800.

Tolle, T.R. et al. (1999) Region-specific encoding of sensory and affective components of pain in the human brain: a positron emission tomography correlation analysis. *Annals of Neurology*, **45**, 40–47.

Toshinai, K. et al. (2003) Ghrelin-induced food intake is mediated via the orexin pathway. *Endocrinology*, **144**, 1506–1512.

Trulson, M.E. and Jacobs, B.L. (1979) Raphé unit activity in freely moving cats: correlation with level of behavioral arousal. *Brain Research*, **163**, 135–150.

Tucker, D.M. and Frederick, S.L. (1989) Emotion and brain lateralization. In H. Wagner and A. Manstead (eds) *Handbook of Social Psychophysiology*. Chichester: Wiley.

Ujike, H. et al. (2002) CNR1, central cannabinoid receptor gene, associated with susceptibility to hebephrenic schizophrenia. *Molecular Psychiatry*, **7**, 515–518.

Ungerleider, L.G. and Mishkin, M. (1982) Two cortical visual systems. In D.J. Ingle, M.A. Goodale and R.J.W. Mansfield (eds) *Analysis of Visual Behavior*. Cambridge, MA: MIT Press.

Van Bemmel, A.L. (1997) The link between sleep and depression: the effects of antidepressants on EEG sleep. *Journal of Psychosomatic Research*, **42**, 555–564.

van de Poll, N.E., Taminiau, M.S., Endert, E. and Louwerse, A.L. (1988) Gonadal steroid influence upon sexual and aggressive behavior of female rats. *International Journal of Neuroscience*, **41**, 271–286.

van Goozen, S.H.M., Frijda, N.H. and van de Poll, N.E. (1995) Anger and aggression during role-playing: gender differences between hormonally treated male and female transsexuals and controls. *Aggressive Behavior*, **21**, 257–273.

Verbalis, J.G. (1991) Inhibitory controls of drinking: satiation of thirst. In D.J. Ramsay and D.A. Booth (eds) *Thirst: Physiological and Psychological Aspects*. London: Springer–Verlag.

Versiani, M. et al. (2002) Reboxetine, a selective norepinephrine reuptake inhibitor, is an effective and well-tolerated treatment for panic disorder. *Journal of Clinical Psychiatry*, **63**, 31–37.

Vertes, R.M. (1983) Brainstem control of the events of REM sleep. *Progress in Neurobiology*, **22**, 241–288.

Videbech, P. (2000) PET measurements of brain glucose metabolism and blood flow in major depressive disorder: a critical review. *Acta Psychiatrica Scandinavica*, **101**, 11–20.

Vink, T. et al. (2001) Association between an agouti-related protein gene polymorphism and anorexia nervosa. *Molecular Psychiatry*, **6**, 325–328.

Vitiello, B., Behar, D., Hunt, J., Stoff, D. and Ricciuti, A. (1990) Subtyping aggression in children and adolescence. *Journal of Neuropsychiatry*, **2**, 189–192.

Vogel, G.W., Vogel, F., McAlbee, R.S. and Thurmond, A.J. (1980) Improvement of depression by REM sleep deprivation: new findings and a theory. *Archives of General Psychiatry*, **37**, 247–253.

vom Saal, F.S. (1983) Models of early hormone effects on intrasex aggression in mice. In B.B. Svare (ed) *Hormones and Aggressive Behavior*. New York: Plenum.

Wagner, H. (1989) The peripheral physiological differentiation of emotions. In H. Wagner and A. Manstead (eds) *Handbook of Social Psychophysiology*. Chichester: Wiley.

Wangensteen, O.H. and Carlson, A.J. (1931) Hunger sensations in a patient after total gastrectomy. *Proceedings of the Society for Experimental Biology*, **28**, 545–547.

Webb, W.B. (1978) The sleep of conjoined twins. *Sleep*, **1**, 205–211.

Webb, W.B. and Agnew, H.W. (1967) Sleep cycling within the twenty-four hour period. *Journal of Experimental Psychology*, **74**, 167–169.

Weinberger, D.R. et al. (2001) Prefrontal neurons and the genetics of schizophrenia. *Biological Psychiatry*, **50**, 825–844.

Weingarten, H.P. (1983) Conditioned cues elicit feeding in sated rats: a role for learning in mean initiation. *Science*, **220**, 431–432.

Weissman, M.M. (1993) Family genetic studies of panic disorder. *Journal of Psychiatric Research*, **27**, 69–78.

Weissman, M.M. et al. (1991) Affective disorders. In L.N. Robins and D.A. Regier (eds) *Psychiatric

Disorders in America: The Epidemiologic Catchment Area Study. New York: Free Press.

Welsh, D.K., Logothetis, D.E., Meister, M. and Reppert, S.M. (1995) Individual neurons dissociated from rat suprachiasmatic nucleus express independently phased circadian firing rhythms. *Neuron*, **14**, 697–706.

Wever, R.A. (1979) *The Circadian System of Man*. New York: Springer–Verlag.

Williamson, E.M. and Evans, F.J. (2000) Cannabinoids in clinical practice. *Drugs*, **60**, 1303–1314.

Wilson, H.C. (1992) A critical review of menstrual synchrony research. *Psychoneuroendocrinology*, **17**, 565–591.

Wilson, W.H., Ellinwood, E.H., Mathew, R.J. and Johnson, K. (1994) Effects of marijuana on performance of a computerized cognitive–neuromotor test battery. *Psychiatry Research*, **51**, 115–125.

Woods, S.C. (1995) Insulin and the brain: a mutual dependency. *Progress in Psychobiology*, **16**, 53–81.

Wu, J.C. and Bunney, W.E. (1990) The biological basis of an antidepressant response to sleep deprivation and relapse: review and hypothesis. *American Journal of Psychiatry*, **147**, 14–21.

Yeo, C.H., Hardiman, M.J. and Glickstein, M. (1984) Discrete lesions of the cerebellar cortex abolish classically conditioned nictitating membrane response of the rabbit. *Behavioral Brain Research*, **13**, 261–266

Yeomans, M.R. and Gray, R.W. (2002) Opiod peptides and the control of human ingestive behaviour. *Neuroscience and Biobehavioral Reviews*, **26**, 713–728.

Yoshidayoneda, E., Tache, Y., Kosoyan, H.P. and Wei, J.Y. (1994) Peripheral bombesin decreases gastric vagal efferent activity in part through vagal pathways in rats. *American Journal of Physiology*, **266**, R1868–R1875.

Young, W.G. and Deutsch, J.A. (1980) Intragastric pressure and receptive relaxation in the rat. *Physiology and Behavior*, **25**, 973–975.

Zald, D.H., Lee, J.T., Fluegel, K.W. and Pardo, J.V. (1998) Aversive gustatory stimulation activates limbic circuits in humans. *Brain*, **121**, 1143–1154.

Zellner, D.A., Rozin, P., Aron, M. and Kulish, D. (1983) Conditioned enhancement of humans' liking for flavors paired with sweetness. *Learning and Motivation*, **14**, 338–350.

FURTHER READING

Section A

Alcock, J. (2001) *The Triumph of Sociobiology*. Oxford/New York: Oxford University Press.

Carey, G. (2002) *Human Genetics for the Social Sciences*. Thousand Oaks, CA: Sage.

Carlson, N.R. (2003) *Physiology of Behavior*, 8th edn. Needham Heights, MA: Allyn and Bacon. (Chapter 5.)

Section B

Carlson, N.R. (2003) *Physiology of Behavior*, 8th edn. Needham Heights, MA: Allyn and Bacon. (Chapter 2.)

Kolb, B. and Wishaw, I.Q. (2001) *An Introduction to Brain and Behavior*. New York: Worth. (pp. 112–164.)

Wilson, J.F. (2003) *Biological Foundations of Human Behavior*. Belmont, CA: Wadsworth. (pp. 27–66.)

Section C

Pinel, J.P. (2003) *Biopsychology*, 5th edn. Needham Heights, MA: Allyn and Bacon. (Chapters 7, 10 and 16.)

Springer, S.P. and Deutsch, G. (1997) *Left Brain, Right Brain*, 5th edn. New York: W.H. Freeman & Co.

Wilson, J.F. (2003) *Biological Foundations of Human Behavior*. Belmont, CA: Wadsworth. (pp. 88–117 and 485–497.)

Section D

Carlson, N.R. (2003) *Physiology of Behavior*, 8th edn. Needham Heights, MA: Allyn and Bacon. (Chapter 3.)

Smock, T.K. (1999) *Physiological Psychology*. Upper Saddle River, NJ: Prentice-Hall. (pp. 122–133 and 146–167.)

Wilson, J.F. (2003) *Biological Foundations of Human Behavior*. Belmont, CA: Wadsworth. (pp. 108–114.)

Section E

Porterfield, S.P. (2000) *Endocrine Physiology*, 2nd edn. St Louis, MO: Mosby.

Rosenzweig, M.R., Breedlove, S.M. and Leiman, A.L. (2002) *Biological Psychology*. Sunderland, MA: Sinauer. (Chapter 5.)

Silber, K.P. (1999) *The Physiological Basis of Behaviour*. London: Routledge. (Chapter 5.)

Section F

Cooper, J.R., Bloom, F.E. and Roth, R.H. (2002) *The Biochemical Basis of Neuropharmacology*, 8th edn. Oxford/New York: Oxford University Press.

Grilly, D.M. (2001) *Drugs and Human Behavior*, 4th edn. Needham Heights, MA: Allyn and Bacon.

Levinthal, C.F. (2001) *Drugs, Behavior, and Modern Society*, 3rd edn. Needham Heights, MA: Allyn and Bacon.

Section G

Kalat, J.W. (2001) *Biological Psychology*, 7th edn; with Infotrac. Belmont, CA: Wadsworth. (Chapter 6.)

Kolb, B. and Wishaw, I.Q. (2001) *An Introduction to Brain and Behavior*. New York: Worth. (Chapter 8.)

Rosenzweig, M.R., Breedlove, S.M. and Leiman, A.L. (2002) *Biological Psychology*. Sinauer. (Chapter 10.)

Section H

Kolb, B. and Wishaw, I.Q. (2001) *An Introduction to Brain and Behavior*. New York: Worth. (pp. 318–335.)

Moller, A.R. (2000) *Hearing: Its Physiology and Pathophysiology*. San Diego, CA: Academic Press.

Yost, W.A. (2000) *Fundamentals of Hearing: An Introduction*. San Diego, CA: Academic Press.

Section I

Breslin, P.A.S. (2001) Human gustation and flavour. *Flavour and Fragrance Journal*, **16**, 439–456.

Doty, R.L. (2001) Olfaction. *Annual Review of Psychology*, **52**, 423–452.

Kruger, L. (ed) (1996) *Pain and Touch*. San Diego, CA: Academic Press.

Section J

Carlson, N.R. (2003) *Physiology of Behavior*, 8th edn. Needham Heights, MA: Allyn and Bacon. (Chapter 8.)

Rowell, L.B. and Shepherd, J.T. (eds) (1996) *Handbook of Physiology*: Section 12. Exercise: Regulation and Integration of Multiple Systems. Oxford/New York: Oxford University Press.

Section K

Hobson, J.A. (1989) *Sleep*. New York: Scientific American Library.

Pressman, M.R. and Orr, W.C. (eds) (2000) *Understanding Sleep: The Evaluation and Treatment of Sleep Disorders*. APA Books.

Refinetti, R. (2000) *Circadian Physiology*. Boca Raton, FL: CRC Press.

Shneerson, J.M. (2000) *Handbook of Sleep Medicine*. Blackwell Science.

Section L

Blessing, W.W. (1997) *The Lower Brainstem and Bodily Homeostasis*. Oxford: Oxford University Press.

Fitzsimons, J.T. (1998) Angiotensin, thirst and sodium appetite. *Physiological Reviews*, **78**, 583–686.

Stricker, E.M. and Sved, A.F. (2000) Thirst. *Nutrition*, **16**, 821–826.

Section M

Ahima, R.S. and Flier, J.S. (2000) Leptin. *Annual Review of Physiology*, **62**, 415–457.

Berthoud, H.R. (2002) Multiple neural systems controlling food intake and body weight. *Neuroscience and Biobehavioral Reviews*, **26**, 393–428.

Raynor, H.A. and Epstein, L.H. (2001) Dietary variety, energy regulation, and obesity. *Psychological Bulletin,* **127**, 325–341.

Schulkin, J. (2003) *Rethinking Homeostasis: Allostatic Regulation in Physiology and Pathophysiology.* Cambridge, MA: MIT Press.

Section N

Abramson, P.R. and Pinkerton, S.D. (eds) (1995) *Sexual Nature: Sexual Culture.* Chicago, IL: Chicago University Press.

De Vries, G.J. and Boyle, P.A. (1998) Double duty for sex differences in the brain. *Behavioural Brain Research,* **92**, 205–213.

McKnight, J. (1997) *Straight Science: Homosexuality, Evolution and Adaptation.* London/New York: Routledge.

Ridley, M. (1994) *The Red Queen: Sex and the Evolution of Human Nature.* Harmondsworth: Penguin Books.

Section O

Davidson, R.J., Scherer, K.R. and Goldsmith, H.H. (eds) (2003) *Handbook of Affective Sciences.* Oxford/New York: Oxford University Press.

Evans, P., Hucklebridge, F. and Clow, A. (2000) *Mind, Immunity and Health: The Science of Psychoneuro-immunology.* London: Free Association Books.

Macmillan, M. (2000) *An Odd Kind of Fame: Stories of Phineas Gage.* Cambridge, MA: MIT Press.

Rolls, E.T. (1999) *The Brain and Emotion.* Oxford/New York: Oxford University Press.

Section P

Dudai, Y. (1989) The Neurobiology of Memory: Concepts, Findings, Trends. Oxford University Press.

Parkin, A. (1999) Memory and Amnesia. Psychology Press.

Sweatt, J.D. and Bicher, D. (2003) Mechanisms of Memory. Academic Press.

Section Q

Obler, L.K. and Gjerlow, K. (1999) *Language and the Brain.* Cambridge/New York: Cambridge University Press.

Section R

Hadaya, R.J. (1996) *Understanding Biological Psychiatry.* New York: W.W. Norton & Co.

Panksepp, J. (ed) (2003) *Textbook of Biological Psychiatry.* New York: John Wiley & Son.

INDEX